ATLAS OF THE PLANETS

ATLAS OF THE PLANETS

PAUL DOHERTY

INTRODUCED BY PATRICK MOORE

McGRAW-HILL BOOK COMPANY

New York · St. Louis · San Francisco · Toronto

This book is dedicated to the memory of my father

ACKNOWLEDGEMENTS

I would sincerely like to thank certain people for their help in the preparation of this book, help without which the book would not have been possible.

In particular my thanks are extended to Patrick Moore for checking the manuscript, making invaluable comments and for supplying the introduction, to Patricia Wood, Producer of 'The Sky at Night' for permission to include the illustrations on pages 127 and 132; to the *Congleton Chronicle* for permission to reproduce the picture on the jacket flap and page 10. Finally to my wife for her patience in putting up with all that is involved in writing a book.

Paul Doherty

Devised and created by
Berkeley Publishers Ltd,
9 Warwick Court, London WC1R 5DJ

Published in United States by
McGraw-Hill Book Company
1221 Avenue of the Americas
New York, N.Y. 10020

First published in United Kingdom by
The Hamlyn Publishing Group Limited.

Library of Congress Cataloging in Publication Data

Doherty, Paul
Atlas of the Planets

Includes: Index . 1 . Planets . 2 . Solar System.
I . Moore, Patrick . II . Title .
G1000.D6 1980 523.4′9 80–12347
ISBN 0–07–017341–9

Phototypeset by Tradespools Ltd., Frome, Somerset.

Printed in Spain.

Contents

NOTE TO READERS

Captions to illustrations drawn or painted from observation will include the following ordered data: date of observation (in figures), time expressed as universal time (UT), type of instrument used (e.g. reflector telescope) and magnification.
In cases where the planet depicted is the main subject of the chapter its name has been omitted from the caption. For example, the illustration of Mercury on page 25, included in the chapter on Mercury, has for the first line of its caption: 22.IV.76 19.05 UT 254mm Refl × 300.

Introduction

Long ago, before astronomy became a science, the Earth was believed to be flat. The stars were points of light, probably attached to an invisible crystal sphere, while the entire heavens revolved round the Earth once a day. The star-patterns were to all intents and purposes fixed, remaining unaltered over periods of many lifetimes, but there were a few luminous points which behaved differently; they wandered around from one constellation to another, and became known as the planets. Five—those which we call Mercury, Venus, Mars, Jupiter and Saturn—were familiar objects even to primitive man.

In Greek times, great progress was made. The movements of the planets were worked out in detail, and a proper 'world system' was drawn up. Admittedly, the Greeks (or most of them) made the fundamental error of retaining the Earth in its proud central position, but at least they realized that the world is a globe, and there were even a few particularly enlightened thinkers who demoted it to the status of an ordinary planet orbiting the Sun. Aristarchus, around 280 BC, anticipated Copernicus by well over a thousand years.

The telescope

Yet little could be learned about the planets themselves before the arrival of the telescope. So far as we know, the first telescopic observations of the celestial bodies were made in or about 1609, and Galileo's classic series of observations began in the first days of 1610. At once, whole new fields of research were opened. Instead of being mere points, the planets became discs, and each showed its own special characteristics, as Galileo was quick to point out. He also used the phases of Venus to demonstrate that the old theories of the Solar System were wrong. Equally significant was his discovery that Jupiter is attended by four moons or satellites of its own, so that in the Solar System there were at least two centres of motion. The Church was not pleased, and everyone will know how Galileo was finally forced into a completely hollow recantation of his heretical view that the Earth orbits the Sun; but by the end of the 17th Century the old geocentric theory had been definitely and permanently laid to rest.

Galileo's telescopes were painfully weak judged by modern standards, and were not nearly so good as present-day binoculars. For instance, they were incapable of showing Saturn's rings in recognizable form, though they could reveal that there was something odd about the shape of the planet. It is therefore rather amazing that Galileo was able to discover the phases not only of Venus (which are, of course, obvious enough; a few keen-sighted people can see them with the naked eye when Venus is in the crescent stage), but also of Mars, which can sometimes appear in the shape of the Moon a few days away from full. At least a start had been made, and as telescopes of better performance were produced, more and more details were revealed. The rings of Saturn were accurately described by Christiaan Huygens less than fifteen years after Galileo's death. The belts and even the Red Spot on Jupiter were discovered, and in 1659 Huygens made the first sketch of a recognizable feature (the Syrtis Major) on the surface of Mars. The age of planetary exploration was well under way.

The first really large and effective telescopes were built by William Herschel, in the latter part of the 18th Century. Herschel, an emigré Hanoverian musician, rocketed to fame in 1781 with his discovery of the planet Uranus. On that occasion he was using a home-made reflecting telescope of modest size, but later he completed a giant with a 1.24-m mirror, and with this he discovered several planetary satellites (Mimas and Enceladus in Saturn's family, Titania and Oberon in that of Uranus). However, the real pioneer of planetary observation was Herschel's contemporary, Johann Hieronymus Schröter, chief magistrate of the little town of Lilienthal, near Bremen. Schröter made many hundreds of planetary drawings, mainly with a telescope built by Herschel, and it may be said that he laid the foundations for future work.

So far as Mars was concerned, the mantle of Schröter fell on the shoulders of two of his countrymen, Wilhelm Beer and Johann von Mädler. From Beer's private observatory in Berlin, and using a 100-mm Fraunhofer refractor, they produced the first reasonable map of Mars, based on work carried out between the late 1820s and the end of the 1830s. But a 100-mm telescope has its limitations, and it was not until later in the 19th Century that really powerful instruments were turned towards the planets. In 1890, there came the arrival of the British Astronomical Association, which has always maintained an observational reputation second to none. The reports of its various planetary sections are full of information, and made an almost complete record which is of unique value today.

Photography takes over

By the turn of the century, astronomy was changing. Sheer visual observation was being superseded by photography, and the age of large refractors was drawing to a close. Reflecting telescopes took over, and for almost all branches of research the photographic plate replaced the human eye. Star-charts became photographically based; essential work, such as the measurement of stellar parallaxes, was handed over entirely to photography. Spectroscopy, too, passed beyond the realm of 'the man looking through the eyepiece'. And with the completion of the 2.5-m Hooker reflector at Mount Wilson, in 1917, astronomy entered a new phase. For the first time it became possible to make exhaustive observations of objects so remote that their light took many millions of years to reach us, and it was by making use of the immense light-grasp of the Hooker reflector, soon after the end of World War I, that Edwin Hubble was able to give conclusive proof that our star-system or Galaxy is but one of many. For instance, the Andromeda Spiral, dimly visible to the naked eye as a blurred patch, is a galaxy considerably larger than ours, and lying at a distance of over two million light-years.

Today, even photography is being to a great extent superseded by electronic devices. But for a moment let us look away from the depths of the universe, and return to the one region in which the eye is superior to the ordinary photographic plate: the region of the planets.

Remember that the planets are

much further away than the Moon, even though they are very close indeed on the cosmic scale. The mean distance between the Moon and the Earth is less than a quarter of a million miles, which is why even a low-power magnification will show so much detail upon its surface. Venus, the closest of the planets, is always at least a hundred times as remote, and is in any case permanently veiled by its dense, all-obscuring atmosphere. Mars never comes much within 56 million kilometres of us, and though its atmosphere is (normally) transparent it is by no means easy to study in detail from such a distance. It has been said, with accuracy, that no Earth-built telescope will show Mars any better than the Moon can be seen with weak opera-glasses; and the remaining planets are further away still.

One might imagine that to study the planetary surfaces, the best procedure would be to use photography with the largest telescopes available. Unfortunately, this simply does not work nearly so well as might be thought, simply because of the effects of our dirty, unsteady atmosphere. When photographing a star-field or a remote galaxy, the main requirement is sheer light-grasp. A slight tremor in the seeing will not do much harm, even when the exposure time has to be several hours in length (as used to be the case before the advent of ultra-modern equipment). But try to photograph a planet, and see what happens. Even with a brief exposure of a few seconds, the air will not stay still. There is bound to be a disturbance, and the delicate planetary detail is lost, leaving only the main features. So long as the observer has to remain on the surface of the Earth, there is virtually nothing that he can do to overcome this problem. Building an observatory on top of a mountain, where the air is thinner, is admittedly a help, but it is not a cure.

A vexed question

To show what I mean by this, consider the vexed question of the canals of Mars. As Paul Doherty explains in his text, the first detailed description of these features was given in 1877 by Giovanni Schiaparelli, using an excellent 225-mm refractor under good conditions (his observatory was situated in Milan, which was much less smoky and polluted then than it is now). For some years, nobody else could see the canal network, but then, in 1886, two French astronomers, Perrotin and Thollon, did so, using the great refractor at the Observatory of Nice. Subsequently, canals became all the rage. Maps of Mars began to look very peculiar indeed, and the network took on an aspect remarkably like that of a spider's web.

The author with Patrick Moore filming 'The Sky at Night'.

Not all astronomers agreed with Schiaparelli and his followers. For instance, Edward Emerson Barnard, noted for his keen eyesight, totally failed to see anything which looked like a canal. The whole problem was bedevilled by somewhat wild theories, due largely to Percival Lowell, to the effect that the canals were artificial, and had been constructed by the local inhabitants as a planet-wide irrigation system. Everything hinged on the reality, or otherwise, of the well-defined canal network, and—and this is the point—photography could not help. Pictures were taken with the world's most powerful instruments, including the Hooker 2.5 m, but were simply not clear enough to solve the mystery one way or the other. Right up to 1965 there were still many reputable supporters of the canal network, even if Lowell's brilliant-brained Martians had been relegated to myth. It was only with the flights of the Mariner and Viking space-probes that astronomers were able to prove, without a shadow of doubt, that the canals of Mars do not exist.

The Space Age

The Space Age began on October 4, 1957, with the flight of Russia's Sputnik 1. The first successful automatic planetary probe was Mariner 2, which by-passed Venus in 1962 and revolutionized all our ideas about that hostile and somewhat sinister world. It was Mariner 2, therefore, which ushered in the new era of planetary exploration. Before that, our main knowledge had been drawn from the work of 'old-fashioned' astronomers who actually looked through their telescopes instead of using them as giant cameras.

Because some of the planets can become so brilliant, it is tempting to think that they are easy objects to study with an adequate telescope. In fact, nothing could be further from the truth. It is childishly easy to 'look', but much more difficult to 'observe' in a properly scientific manner. To carry out useful work, there are various essentials:

Patience Sporadic observations, carried out every now and then, are of very little scientific use. It will always happen that a particularly interesting feature will be missed simply because the would-be observer is not on the alert. It is also true that the enthusiastic observer must reconcile himself to keeping what may be termed 'uncivilized hours'. For example, making a series of useful observations of Mars before the planet has reached opposition means being at the telescope during the hours immediately before dawn.

Experience The more one practises, the more one will see. I may perhaps cite a personal experience here. Years ago, a teenage enthusiast came to see me, and asked whether he could use my telescope (a 400-mm reflector). For some weeks I was saying to him, 'What! can't you even see *that*?' when referring to some feature on Jupiter or Mars, which happened to be the two planets on view at that time. With my years of experience, and my trained eye, I could see more detail than he could even suspect. I think that it was about three months before we could see an equal amount, but within a year the whole situation was reversed; he could record details which were completely beyond me. He had become a better observer, and he has remained so, but he had to go through the apprenticeship of learning, and there is no short cut.

Visual acuity Here, of course, one is at the mercy of Nature, and one can do nothing about the quality of one's eyesight. It is painfully obvious that some people are more keen-sighted than others. I had a reminder of this in 1978, when Paul Doherty was staying with me at my home in Selsey. It was early afternoon, and we decided to have a look at Venus, which was high in the sky. I was about to go into the dome and set the circles, in order to swing the telescope on to Venus, when Paul said 'Don't bother—I can see it!' I was frankly incredulous. The sky was bright, and I was very dubious as to whether even a Lynceus could have detected Venus under such conditions. I was silenced when Paul went to the telescope, swung the dome, looked through the finder, and—there was the planet.

Eyesight of this sort is a gift. But I repeat that experience is a great leveller. If you are not keen-sighted, do not despair—make the most of the acuity you have. Incidentally, I have found that most spectacle-wearers do better by taking their glasses off and adjusting the telescope focus to compensate, though I agree that this is not an invariable rule.

Draftsmanship Here we have again to reckon with varying degrees of ability, and here again the observer must reconcile himself to making the best of whatever gifts he has.

Drawing the planets

Perhaps I may be forgiven for citing my own case, because I think that it is relevant in the present context. To me, drawing in perspective is a total impossibility. I have no idea of how to do it, and if I lived to be a hundred years old I would never learn. Admittedly, there is no real 3-D effect in a telescopic view, but it is undeniable that I am completely inartistic, and no drawing that I make will look like 'the real thing'. Yet this is not totally disastrous. Within my marked limitations, I can make a sketch of a planet which may not be beautiful, but which is an accurate record. And scientifically this is the most important thing.

My own method, therefore, is to go to the telescope, spend some time in surveying the object to be drawn, and then make a sketch as quickly as I can. (With most planets one must not take too long about it, or there will be awkward shifts in the visible features due to the planet's axial rotation. Check carefully.) I then go to the desk in the observatory and make a 'fair copy' of the sketch, drawing it as faithfully and artistically as I can, but making the accurate positioning of features, with their intensities, of paramount importance. I next return to the telescope and re-check the drawing, making allowance for the rotational shift of the planet in the interim. Take care, of course, to add details such as date, time, telescope type and aperture, seeing conditions and name of observer. (Like most or all observers, I use the seeing scale devised years ago by E. M. Antoniadi, Greek by birth but French by naturalization. In this scale, I denotes perfect seeing; II, III and IV are progressively worse, and at seeing V one would not normally bother!)

But for an observer who is also artistic, things are quite different, and it is into this category that Paul Doherty comes. Unlike me, he is capable of making his actual drawing at the telescope, merely 'finishing it off' afterwards, and he also has the gift of producing a result which is uncannily like the view obtained through the eye-piece. There is no point in my trying to describe exactly how he does it, but I would stress that anyone who does not totally lack ability would be wise to attempt his methods rather than mine. But—I do not apologize for stressing this—never, repeat *never*, sacrifice accuracy in the cause of artistry. A clumsy sketch which represents the detail faithfully is much more valuable than a beautiful drawing in which the features are misplaced. The ideal is to combine both.

Before discussing the planets in rather more detail, I must say something about telescopic equipment, because this is so often a stumbling-block. It is, alas, true, particularly in these days of rampant inflation, that good telescopes are expensive items. When I was ten years old I bought a fine 75-mm refractor for the sum of £7 10s, and with it I well remember seeing the famous white spot on Saturn discovered by W. T. Hay (Will Hay, the great comedian of stage and screen!) as well as the belts and moons of Jupiter and the rings of Saturn. This, however, was in 1933. I still have the telescope, but by now its value has risen well over the £100 mark. Moreover, good second-hand telescopes today are about as common as great auks.

Buying a telescope

It is always possible to buy small astronomical telescopes at what seems to be a tolerable cost—a few tens of pounds. Yet these will always be too weak to be of satisfactory performance except for casual viewing of star-fields and lunar craters. In my view, and I must emphasize that this is personal only, it is not sensible to pay a great deal for any telescope of aperture less than 75 mm (for a refractor) or 150 mm (for a reflector). The field of view will be restricted; so will be the light-grasp, and the overall impression will be disappointing, at least insofar as planets are concerned. Recently I tested a professionally-made 100-mm reflector, and turned it towards Jupiter. I could just about see the two main belts, and the four Galilean satellites were easy, but that was all—and it was, frankly, all that could have been reasonably expected.

No. My considered advice is to save up for a telescope of larger aperture, which will be adequate to show real planetary detail. Moreover, ensure that the mounting is firm, as an instrument mounted upon a stand with the rigidity of a blancmange will be useless by any standards. An equatorial mounting, with clock drive, is highly desirable, though not absolutely necessary (I have carried out a great deal of planetary observation with my 312 mm reflector, which is mounted upon a massive altazimuth and has manual slow motions).

Making a telescope

The other alternative, of course, is to make a telescope for oneself. Generally speaking, constructing a lens for a large refractor is rather too major a problem for the average amateur—though I have known it be done—and a better choice is a reflector. Either buy the optics and mount them, which is a sheer question of mechanics, or else grind the mirror and make it to your own requirements. This is no place to discuss telescope-making, but I must comment that Paul Doherty's telescope, used for virtually all the drawings in this book, was home-made apart from the optics and it works splendidly. I have used it myself, so I can speak with full authority!

There is another point which must be considered, honestly and with complete frankness. Before 1962, it was permissible to say that much of our knowledge of planetary surfaces was derived from visual observations, carried out largely by amateurs with modest equipment. Before and immediately after the war, this was also true of the Moon. But today the space-probes have to a large extent taken over, and our ideas about the planets have been changed beyond all recognition. When I began observing, in the 1930s, who would have thought that Mars was a world of giant volcanoes, or that Mercury was cratered in the same way as the Moon? No Earth-based observers can see detail which is glaringly obvious from relatively close range, and it has even been suggested that these new techniques make amateur planetary observation a pastime 'for amusement only'. I do not agree. Of course no telescopic observer can see the craters of Mercury, the ghostly ring of Jupiter or the scorching rocks of Venus. Actual planetary cartography is to a great extent outmoded, as is also true of the Moon. But the observer can still make himself useful in studying what may be termed time-dependent phenomena, such as the onset and development of dust-storms on Mars or the changing aspects of the transitory features on the gaseous surface of Jupiter. And there is always the chance of making a valuable unexpected discovery, as Will Hay did when he detected Saturn's great white spot in 1933.

Let us now take a brief look at the planets, one by one, and see what we can make of them.

Mercury The first thing to do is to find it, and this is often difficult. Because it is so close to the Sun, and is not a great deal larger than the Moon, it is an elusive object, and it is visible with the naked eye only when low down in the west after sunset or low down in the east before dawn. The best observations of it are made when it is high in the sky, though the Sun will also be high and contrast is correspondingly reduced.

I must make one vital point here. Not long ago I read a book in which it was said that the observer could well 'sweep around' during daytime, using binoculars or a telescope, so as to locate Mercury. I can only assume that the author had taken leave of his senses, because any such procedure is terrifyingly dangerous. Sooner or later the Sun will enter the field of view, with tragic results. To locate Mercury during daylight, you need a telescope with really accurate setting circles, and if you have to sweep at all, always do so *away from the Sun.*

Actually, little will be seen on Mercury apart from the characteristic phase. Paul Doherty's drawings show some vague shadings, but one needs keen sight to make out even as much as that. Before the epic flight of Mariner 10 nothing definite was known about the surface features. Craters seemed logical enough, and so it proved, but no Earth-based telescope is capable of showing them. Search for Mercury by all means, but do not be disappointed at the lack of detail.

Venus Here we have a world of totally different type, and our ideas have somersaulted since the start of the Space Age. Before Mariner 2, it was widely believed that the surface of Venus might be oceanic, in which case primitive life might have existed there. It was even suggested that Venus could be a world in a Pre-Cambrian stage of development. Actually, the surface temperature is not far short of 500 degrees Celsius; the atmosphere is made up chiefly of carbon dioxide; the surface pressure is about 90 times that of the Earth's air at sea-level, and the clouds contain large amounts of sulphuric acid—making Venus just about the most lethal world in the entire Solar System. The Russian probes Veneras 9 and 10, which made controlled landings, sent back pictures showing rock-strewn landscapes, and radar measurements from Earth have indicated the presence of large, shallow craters, but telescopes on their own will show little. On many occasions Venus looks quite featureless.

However, shadings are visible sometimes, together with bright regions over what we now know to be the poles of the planet.

Mars No planet can have caused so much controversy as Mars. The canal problem was finally resolved less than twenty years ago, and even now we cannot be sure whether or not traces of life exist there. But certainly the space-probe results have made astronomers revise all their cherished ideas.

Before 1965, it was widely believed that Mars must have a gently undulating surface, with no major mountains or valleys anywhere. The famous dark areas were regarded as depressed regions, probably old seabeds filled with lowly vegetation, and the polar caps were dismissed as being thin layers of hoar-frost less than an inch deep. The atmosphere was believed to have a ground pressure of between 80 and 90 millibars, and to be made up chiefly of nitrogen. The picture we have today is as different as it could be. On Mars there are vast volcanoes, gaping craters and yawning valleys. The dark areas are not basins—indeed some of them, such as the famous Syrtis Major, are lofty plateaux. There are no vegetation tracts, and the residual polar caps contain a large quantity of water ice. Everywhere on the surface the probe pictures show traces of past water activity, and there is no reason to doubt the existence of subcrustal ice. The main disappointment has been the revelation that the atmosphere is made up almost entirely of carbon dioxide, and that the ground pressure is everywhere below 10 millibars, corresponding to what we normally call a laboratory vacuum. No advanced Earth-creature could survive on Mars in such an atmosphere.

The author with his 419-mm reflector telescope, built almost entirely from scrap.

Space probes

Thanks to the orbiting probes (Mariner 9 and the relevant sections of the Vikings), we now have detailed charts of the whole of the Martian surface, and cartography, as such, is therefore complete. But this does not mean that the Earth-based observer has nothing left to do. Mars is not a static, changeless world. Extensive

dust-storms spring up, spread and may cover the whole of the planet in an amazingly short time. The outlines of the dark regions are not absolutely constant, but vary within certain limits, for reasons which are still not properly understood. And, of course, the polar caps show seasonal cycles, while there are also isolated clouds to be seen in the Martian atmosphere. Once again the observer will be wise to concentrate upon these variable phenomena and, despite the Mariners and the Vikings, he can still make useful contributions to knowledge.

Jupiter Here we have a planet whose surface is always changing. It is surely significant that during 1979, when the two Voyager probes were approaching the planet, amateur observers with adequate equipment were called in to maintain a close watch upon the planet, so that the various features could be constantly monitored. One requirement was the determination of the longitudes of different features. Drawings of the surface are also most valuable, because they can be used to trace the life-histories of specific features.

The main problem of making a full-disc drawing of Jupiter is that one does not have much time to spare. The rotation period of the planet is less than 10 hours, so that the features shift across the disc with alarming rapidity. The best method is to put in the main details, leave them unaltered with regard to position, and then fill in the minor features to correspond with the general background. This is what Paul Doherty has done for the drawings given here, and the procedure is certainly to be recommended to observers.

Scientific accuracy

I must digress for a moment to comment briefly upon the use of colour. There can be no doubt that coloured pictures are much more spectacular than those made in black-and-white, and the artistic observer can make them easily enough. In fact, even the non-artist can do so; I, personally, use crayon. But I come back again to the all-important point that artistry must be subordinate to accuracy, and reliable monochrome is preferable to arbitrary attempts at colour, which are never justified.

Satellite phenomena are easily recorded, and a small telescope will show that the four Galileans are definite discs rather than points, but surface details on them are hard to make out even with giant instruments. I have seen some vague features when using very large refractors such as the 825 mm at Meudon, near Paris), but telescopes of 'amateur size' will not show them, and before the space-probe flights we had no real idea of what the satellites were like.

Saturn Here we have a planet which poses serious problems for the non-artist. The ring system is always changing in angle, and using prepared outlines, as I do, may be dismissed as a coward's way out. But there is no doubt that Saturn is the loveliest object in the sky. With an adequate telescope it is a breathtaking sight and a challenge for the artist. Paul Doherty makes his drawings 'freehand' at the telescope. Once, when he joined me in a *Sky at Night* programme recorded from his Stoke observatory, he produced a magnificent drawing at the telescope, finished it off indoors, and completed it in less than an hour. I have it now, suitably framed, and hanging in my study in a place of honour. Not many people can do this. Yet oddly enough, features on Saturn's disc are more important to observe, scientifically, than the glory of the rings. Spots are much less common than those of Jupiter, but they do appear sometimes, and the painstaking amateur has the best chance of detecting them as soon as they appear. Valuable work can also be done with regard to satellite magnitudes. Iapetus, in particular, is notoriously variable.

Telescopic planets For obvious reasons, little can be seen on the pale discs of Uranus or Neptune. Uranus is decidedly greenish, and ill-defined bright and dark bands can be made out under good conditions, but I admit that I have never seen anything definite upon the disc of Neptune, and Pluto appears as nothing more than a faint star. The asteroids, too, appear stellar, which applies even to Ceres, by far the largest of the swarm. All the same, these relatively inconspicuous members of the Sun's family are well worth seeking out.

Finally, I have two more points to make. The first concerns what may be termed imaginative drawings—scenes that would meet the eye of the observer who is standing on another world, or is in orbit around another planet. Many such sketches have been produced in the past. Some have proved to be close to the mark, while others have been, frankly, wide of it. In some cases, past artists cannot be blamed for their inaccuracies. For instance, in pre-Viking days it was thought certain that the sky as seen from Mars would be dark blue, whereas in fact it turned out to be salmon-pink (owing to the large amounts of very fine dust suspended in the thin Martian atmosphere). But today we have information which is really reliable, and the artist can draw upon sound scientific knowledge as well as using his imagination where necessary.

Outstanding observer

In this field Paul Doherty is exceptionally well equipped. For many years now he has been recognized as an outstanding planetary observer who combines accuracy with artistic talent. He is also an artist in the more conventional sense of the term, and his futuristic drawings are as precise as they can possibly be made at the present time. It will, I fear, be a long time before any astronaut ventures as far as remote, chilly Pluto but, when this happens, I feel that the scene will be very much the same as that visualized in Paul Doherty's drawing on page 132 of this book. Of course there will be discrepancies in detail, but nothing basic.

This *Atlas of the Planets* is unique of its kind. Rather than use hackneyed sketches made with huge telescopes, or photographs taken from Earth which are bound to be relatively lacking in detail, Paul Doherty has used his own exceptional skill to provide drawings which are faithful as well as spectacular, and which will be of real scientific use to the observer. Compare these pictures with the views you will obtain with an adequate telescope and you will see what I mean. I have been honoured to write an introduction to a book which has so much to offer.

PATRICK MOORE
Selsey, March 1979.

The planets in general

There are nine major planets in our Solar System and these are objects of great interest to amateur and professional astronomers alike. Of the nine, five are bright objects and can fairly easily be distinguished from the 'fixed stars' even without a telescope. The patterns of the stars change so slowly that no appreciable difference will be noticed by several generations of observers. The planets, on the other hand, do appear to move and although this motion is relatively slow, their position against the background stars will, in some cases, alter noticeably from night to night. In fact, if a planet close to us passes relatively near to a bright star, its changing position will be apparent over a couple of hours. The motions of all the planets will be quite obvious after observing for a few weeks, with the most distant ones appearing to move the least.

Ancient astronomers noticed the movement of these five bright objects, hence the name, 'planets' meaning 'wanderers'. Three of the other planets, being remote and rather faint, were not known to the ancients. The remaining one of the nine is, of course, our own Earth.

The basic motions of the planets, although apparently complex, are easily explained. All of these bodies are in orbit around the Sun but two of them, namely Mercury and Venus, have smaller orbits than the Earth and are closer to the Sun. They complete their orbits in less time than the Earth's orbital period of one year because they do not have so far to travel and orbital speed increases as distance from the Sun decreases. Consequently, both these planets will at certain times be found on the opposite side of the Sun to the Earth and, since the orbits of all the planets lie in roughly the same plane, they will then appear very close to the Sun in the sky. They are not actually close, but appear to be so purely because of their alignment. This is called 'superior conjunction' and is a very unfavourable time to look for them (see figure, page 14).

Since the direction of travel around the Sun is the same for all the planets,

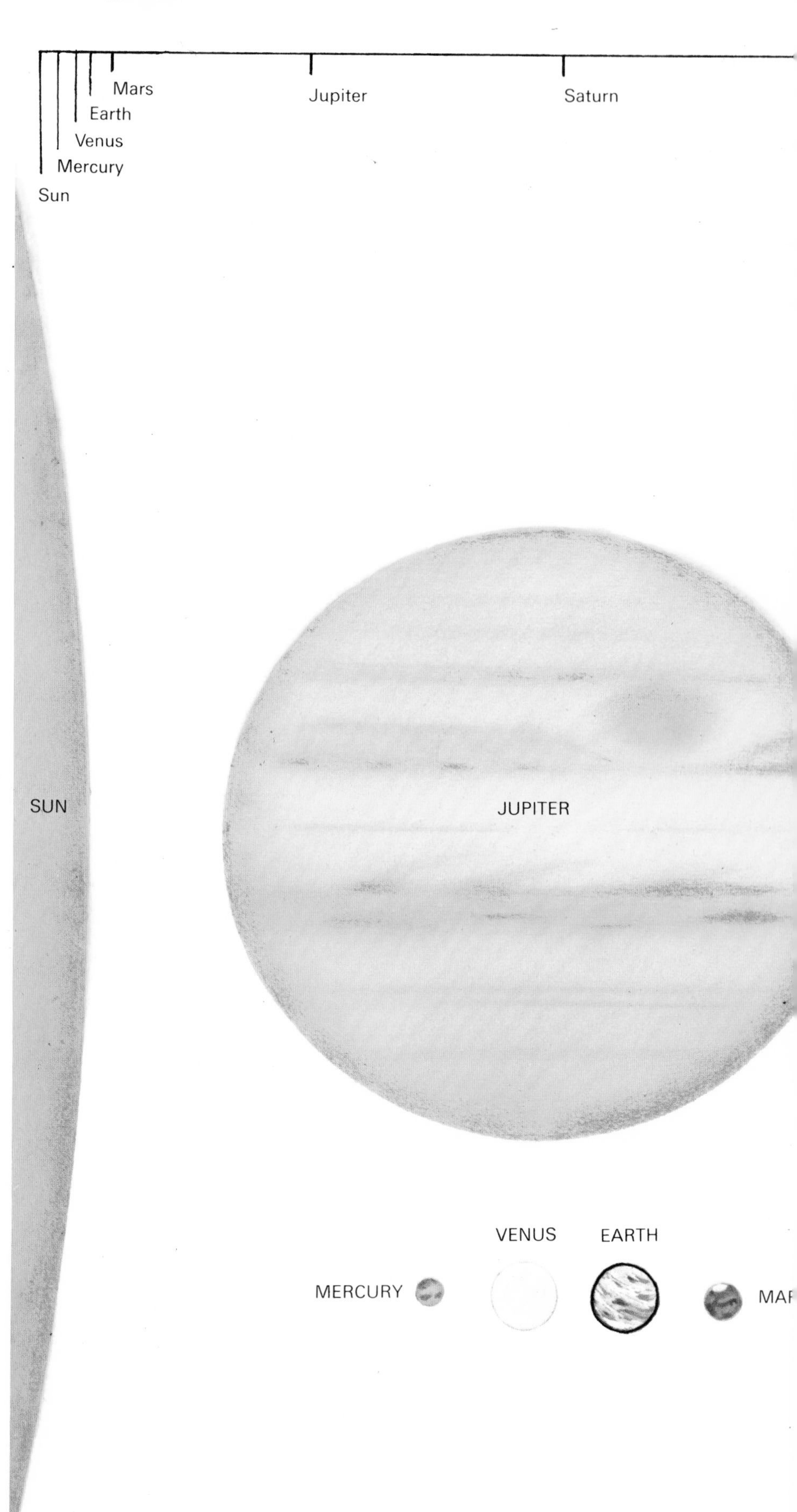

Scale drawing showing the planets' respective sizes and distances from the Sun.

Uranus Neptune Pluto

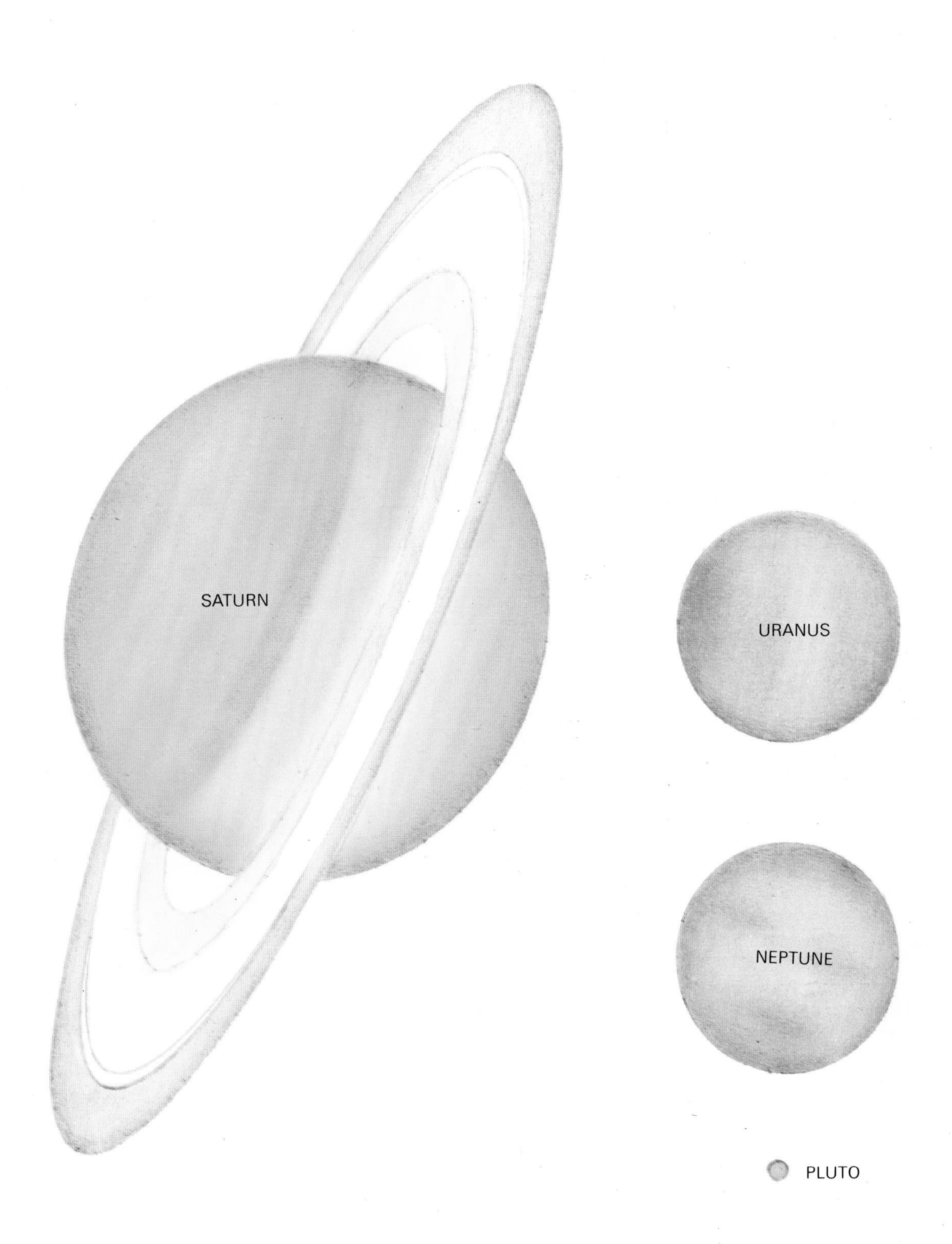

Mercury and Venus, in completing their orbits more quickly than our own, will appear for a time to chase the Earth, eventually catching up with, and passing, it on the inside. When these planets are situated between the Earth and the Sun the phenomenon is called 'inferior conjunction' (see figure). They are then at their closest to the Earth and appear larger than at any other point of their orbit. The term 'inferior' does not refer to size of the planets in this case, although both Mercury and Venus are smaller than the Earth, but to their closer proximity to the Sun.

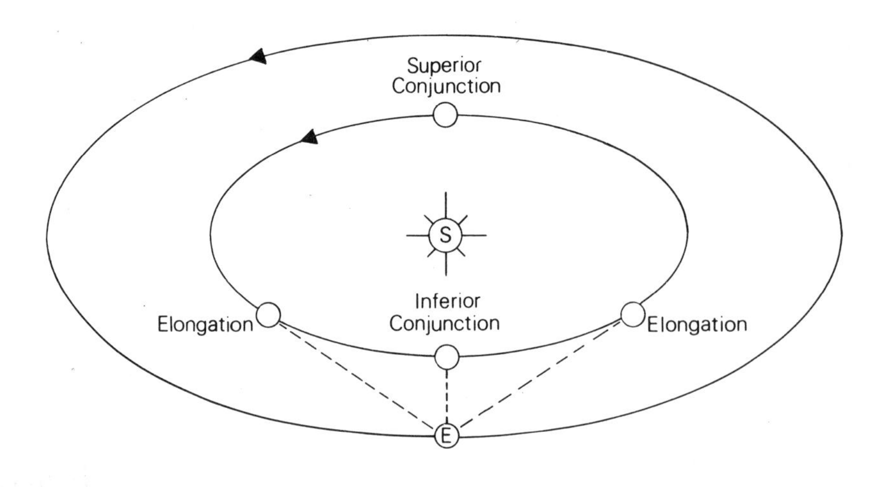

Various positions displayed by the inferior planets with respect to the Sun (S) and Earth (E).

Greatest elongation

During the process of catching up with the Earth, Mercury and Venus will appear to move away from the Sun in an easterly direction. They will be visible in the evening sky and will set in the West only a few hours after the Sun. Eventually a point will be reached where, relative to the Sun, these planets appear stationary. They are, in fact, moving towards the Earth. It is at this point, known as 'greatest elongation', that they attain their greatest angular separation from the Sun. When this occurs in the evening sky it is called the 'greatest eastern elongation'. The planets will then appear to move back towards the Sun to pass through inferior conjunction. After this, they will again move out from the Sun, this time in a westerly direction, and become visible in the morning sky, rising at best only a few hours before the Sun. Stationary points will eventually be reached as the greatest separation is once more attained. This is referred to as the 'greatest western elongation' and the planets are then moving away from the Earth. Finally, they will once again close in on the Sun, pass through superior conjunction and the whole process starts once more.

The conditions for observing Mercury are much less favourable than for Venus on account of Mercury's proximity to the Sun. Mercury can only rise one or two hours before the Sun, and must be near greatest elongation to be visible. It is rarely seen in a dark sky, particularly from the latitudes of the British Isles. Venus, with its greater distance from the Sun, can at times rise or set five hours before or after the Sun, and is often seen in a dark sky.

Retrograde motion

The apparent motions of the outer or superior planets—that is, those orbiting the Sun at greater distances than the Earth—are in some ways similar to the motions of the inner planets, despite certain basic differences. In these cases, Earth is the inferior planet moving more quickly around the Sun, and it is up to Earth to do the chasing. There is a stage

Elongations of an inner planet as seen from Earth with the Sun just below the horizon.

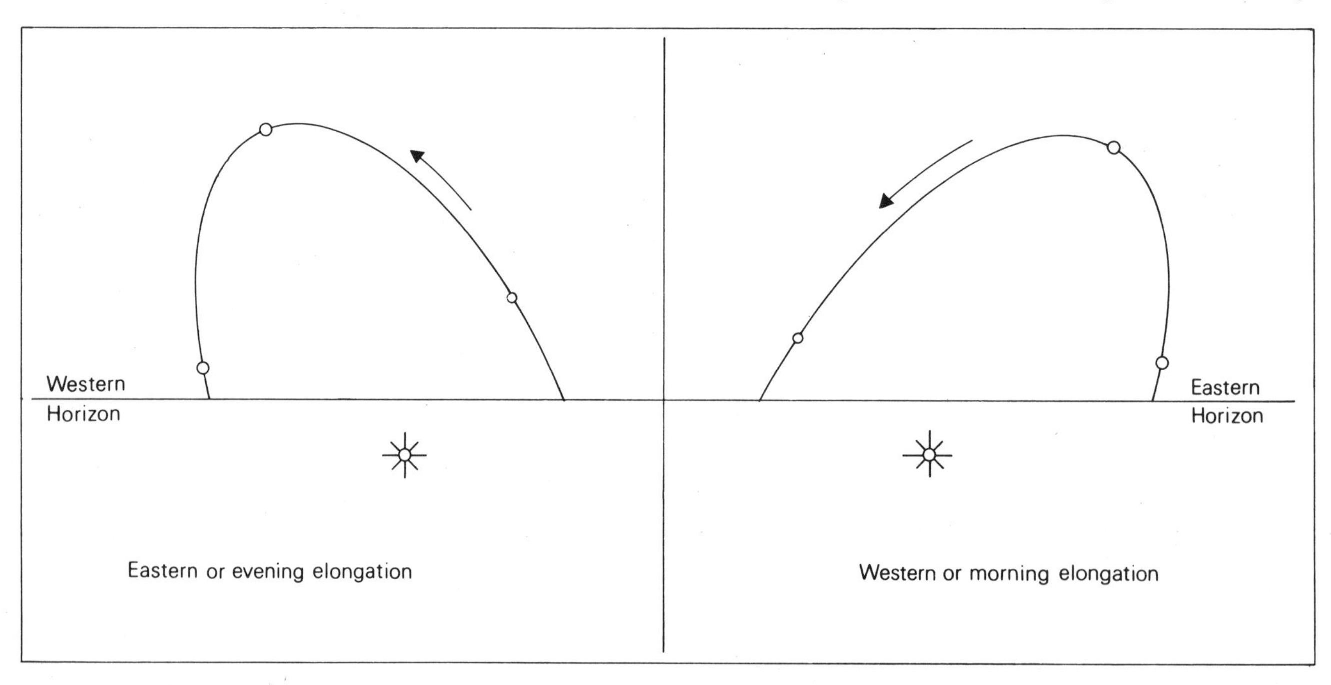

where all outer planets pass, at one time or another, through superior conjunction. Beyond this point, the Earth begins to gain on the outer planet, which will appear to move out from behind the Sun in a westerly direction, first becoming visible in the morning sky. To complicate matters, the motion of the outer planet against the background stars appears eastward. (The impression of westerly motion is only created by the Sun moving eastward at a quicker rate than the outer planet.)

As the Earth then moves to a part of its orbit where it begins to swing in to pass the planet on the inside, the planet, as the Earth approaches, will for a short time appear stationary in relation to the background stars. Just as the Earth is passing the planet, the planet will seem to move in a westward direction against the stars. This is referred to as a 'retrograde motion'. At this time the planet will be seen in a part of the sky which is opposite to the Sun, thus 'standing at opposition'. The planet will then appear on the meridian at midnight, or due south to an observer in the northern hemisphere. This is the most favourable position for observation.

After opposition, the situation is virtually reversed. The planet moves towards the Sun, which now appears to the west, and its motion once more becomes eastward, or direct, against the background stars as Earth leaves it behind. Finally, the planet is lost in the evening twilight prior to superior conjunction and the whole process can begin again.

The length of time this process takes is not only governed by the Earth's orbital or sidereal period, but also by the orbital period of the planet concerned. In the case of Mars, for instance, it is quite a lengthy process. The sidereal period of Mars is 686.9 days; that of the Earth, 365.2 days, or a little more than half that of Mars. As a result, one year after a particular opposition of Mars, the Earth will have returned to the part of its orbit held at the time of this

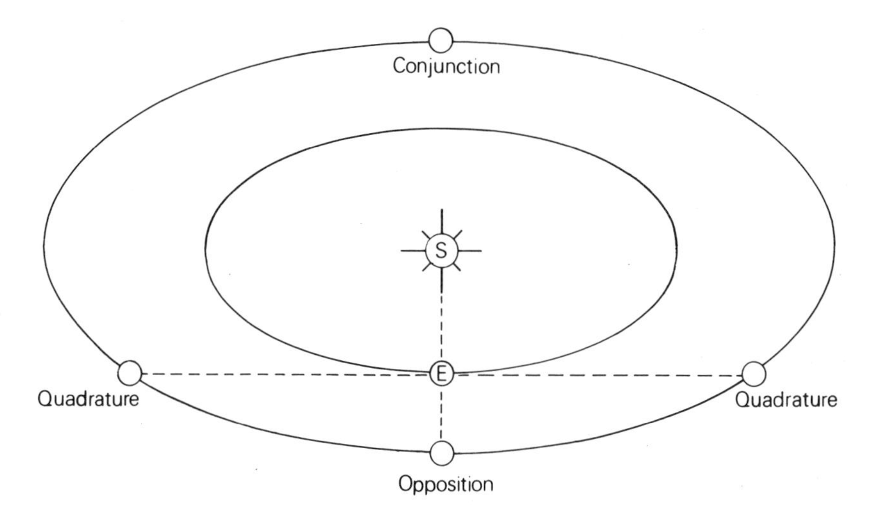

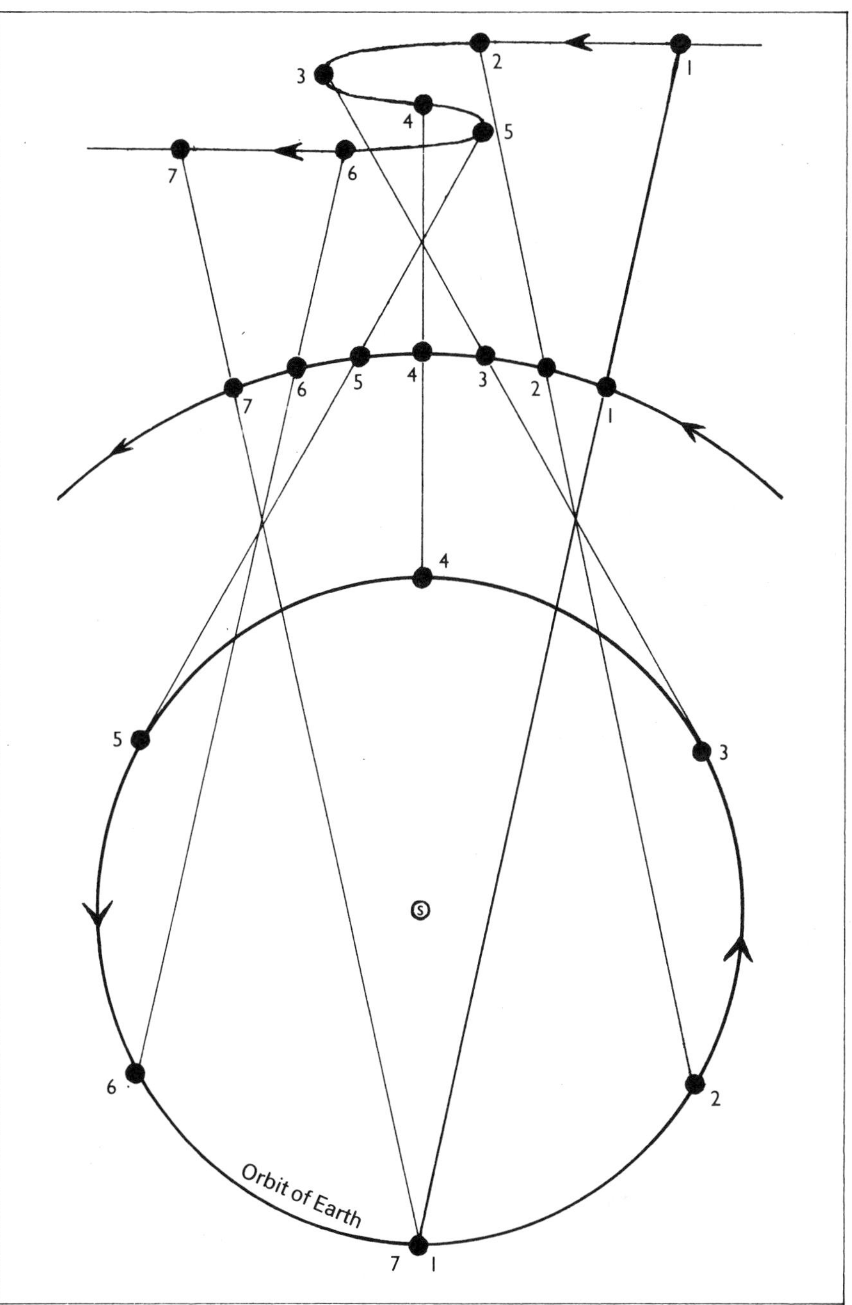

Above *Various positions of superior planets with respect to the Sun (S) and Earth (E).*
Right *The apparent path of an outer planet traced from Earth at seven stages in Earth's orbit.*

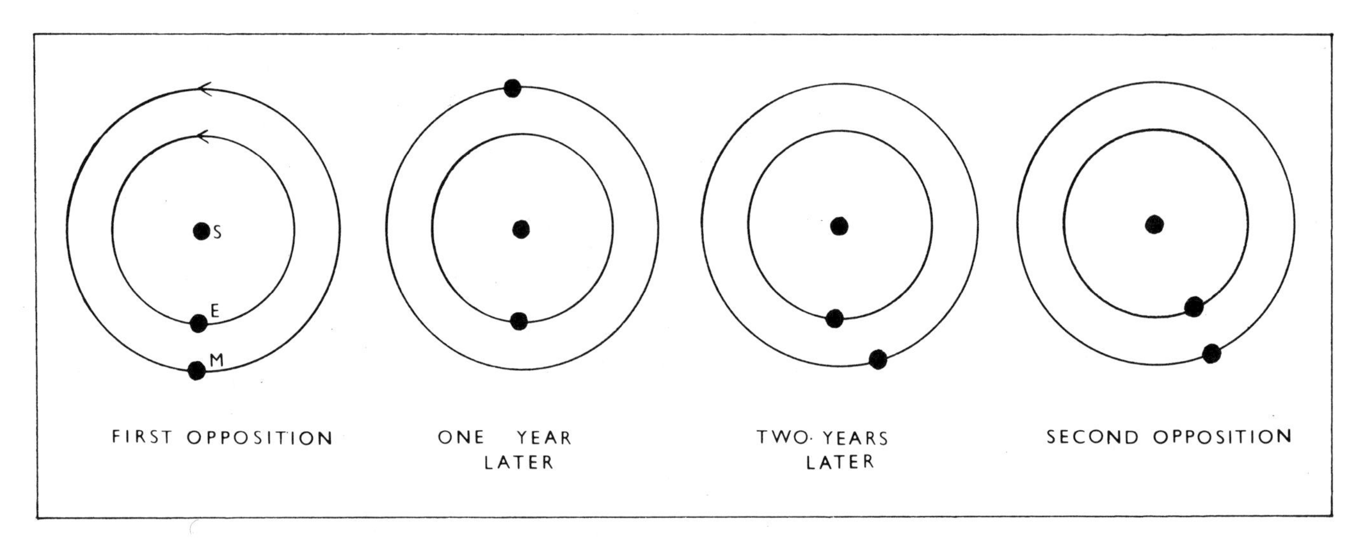

Diagram showing the synodic period of Mars (M).

opposition, while Mars will have completed about half of its orbit and will appear on the opposite side of the Sun, at superior conjunction. A further 14 months are required to bring Mars back to opposition for the second time. The period from one opposition to the next is called the 'synodic period' and for Mars the mean, or average, synodic period is 779.9 days.

For the more remote planets this synodic period becomes shorter, the most distant planet, in fact, having the shortest. Jupiter, for instance, has a sidereal period of slightly more than 11 years 10½ months and will thus complete about one-twelfth of its orbit for each complete orbit of the Earth. Oppositions of Jupiter occur every 1 year and 34 days, or slightly over 13 months. (This is interesting because since there are 13 zodiacal constellations, and Jupiter takes around 12 years to complete one circuit of the ecliptic, it seems to pass from one constellation to the next with each successive opposition.) Saturn, taking 29 years 167 days to complete an orbit, has a synodic period of 1 year 13 days. Uranus, with an orbit taking 84 years and 1 week, is brought into opposition only 4 days later each year. Neptune's sidereal period is about 165 years. Thus, its synodic period is only 1 year and 2 days. Finally, Pluto's 247-year sidereal period results in a synodic period of one day longer than a year.

For the inner planets the synodic period is the time between each successive inferior conjunction. Mercury, with the shortest sidereal period of all, at only 88 days, passes through inferior conjunction several times a year and has a synodic period of 115.9 days. Venus has a sidereal period of 224.7 days and a synodic period of 583.9 days, or about 20 months, which is the second longest after Mars. It is worth noting that the synodic periods for the three planets Mercury, Venus and Mars are longer than their sidereal periods. For those planets outside the orbit of Mars the opposite will be the case, however.

The Zodiac

As mentioned earlier, the orbits of the major planets lie roughly in the same plane. Only Mercury and Pluto have orbits that are highly inclined to the rest, but even so their inclination is not excessive. Thus, all the planets seem to keep to a region of sky forming an imaginary band that encircles the celestrial sphere. This band is about 17 degrees in width and is called the 'Zodiac'. Running along the centre of this band is a line, again imaginary, which is called the 'ecliptic'.

To understand this better, consider the whole sky as an enormous sphere encompassing the Earth. The axis of the Earth, which is the line passing through the poles, if projected on to this sphere indicates the celestial poles. The north celestial pole is easy to find because of the fairly bright nearby star Polaris, or North Star. The equator of the Earth, which may be considered as a plane through the centre of the Earth, perpendicular to the axis, becomes the celestial equator when projected on to the sphere. The ecliptic is an imaginary line on the celestial sphere traced by the path of the Sun. Since we are orbiting the Sun, the Sun will appear to follow this line, completing one circuit each year. The Zodiac itself is a path on the celestial sphere which all the planets follow, with the single exception of Pluto, whose orbit is sufficiently inclined to the plane of the ecliptic to allow it to move occasionally into regions of the sky bordering the Zodiac.

Because the Earth's axis is tilted to the plane of the ecliptic by 23° 27′, the apparent paths of the Sun and planets on the celestial sphere take a rather unusual course, relative to the celestial equator. Perhaps the best way to describe the situation is to consider an observer standing on the Earth's equator. To him the celestial equator would appear as an imaginary line from the eastern horizon to the western horizon and passing directly overhead. The ecliptic, however, is inclined to the celestial equator by 23° 27′, an amount corresponding to the Earth's axial tilt. At certain points, therefore, the ecliptic will appear to the south of the celestial equator and at other points to the north, its maximum separation being, of course, 23° 27′.

In two positions, 180° apart, the ecliptic intersects the celestial equator. These are called the 'vernal', or 'spring', 'equinox' and the 'autumn equinox'. The first of these points is at present found in the constellation of Pisces, the second in Virgo. The point where the ecliptic has its maximum displacement north is called the 'summer solstice'; that where it has its maximum displacement south is the 'winter solstice'. These are situated in the constellations of Gemini and Sagittarius respectively. Altogether there are thirteen constellations strung along the ecliptic, not twelve as is often believed. They are Gemini, Cancer, Leo, Virgo, Libra, Scorpius, Ophiuchus, Sagittarius, Capricornus, Aquarius, Pisces and Taurus. Their signs are, with the exception of Ophiuchus, the well-known signs of the Zodiac.

Observing from Britain

The motions of the planets are observed less easily from the latitudes of the British Isles than from the equator. If an observer on the equator were to shift his position to a northern latitude of, say, 53°, corresponding to the latitude of central England, the celestial equator would appear displaced towards the southern horizon by a corresponding amount and, at its highest point in the south, would attain an altitude of only 37° (37° + 53° = 90°). Since the ecliptic can appear either 23° 27′ south or north of this, depending on the part of the sky visible, its position could be as much as 60° 27′ or as little as 13° 33′ above the southern horizon at a point due South.

A planet situated in the constellation of Sagittarius will be very low in the sky as seen from this country. If, on the other hand, a planet is situated in the constellation of Gemini, it will be very high up when due south. In the former, a planet would be very poorly placed for observation since we would be seeing it through dense layers of the Earth's atmosphere. The brightness and clarity would be very severely affected. With the planet high in the sky, however, it would be viewed through a much thinner layer of atmosphere and so would be brighter and sharper. This is bringing us to the subject of seeing conditions, which will be discussed more fully later in this chapter.

For the outer planets, an opposition occurring during the winter months will place that planet well north of the equator, so that this is the best time for the northern observer. Oppositions occurring during the summer months place the planet well south of the equator and for the northern observer this would not be so good. Spring oppositions imply that a planet is crossing the equator going south, while Autumn oppositions mean that a planet is crossing the equator moving north. For the two inner planets eastern, or evening, elongations in the spring, and morning, or western, elongations in the autumn, place the planet well north of the Sun and conveniently situated for observers in this country. Spring morning and autumn evening elongations place the planet well south of the Sun, so these are not favourable for the latitudes of the British Isles. Astronomers living on the equator are very fortunate so far as the planets are concerned, since the ecliptic is never situated more than 23.5° north or south of the celestial equator and the planets will always pass close to the zenith (point overhead). Observers at the poles would never have a very favourable opposition and on occasions would not see an opposition at all, because the planet concerned would be below the horizon throughout the opposition.

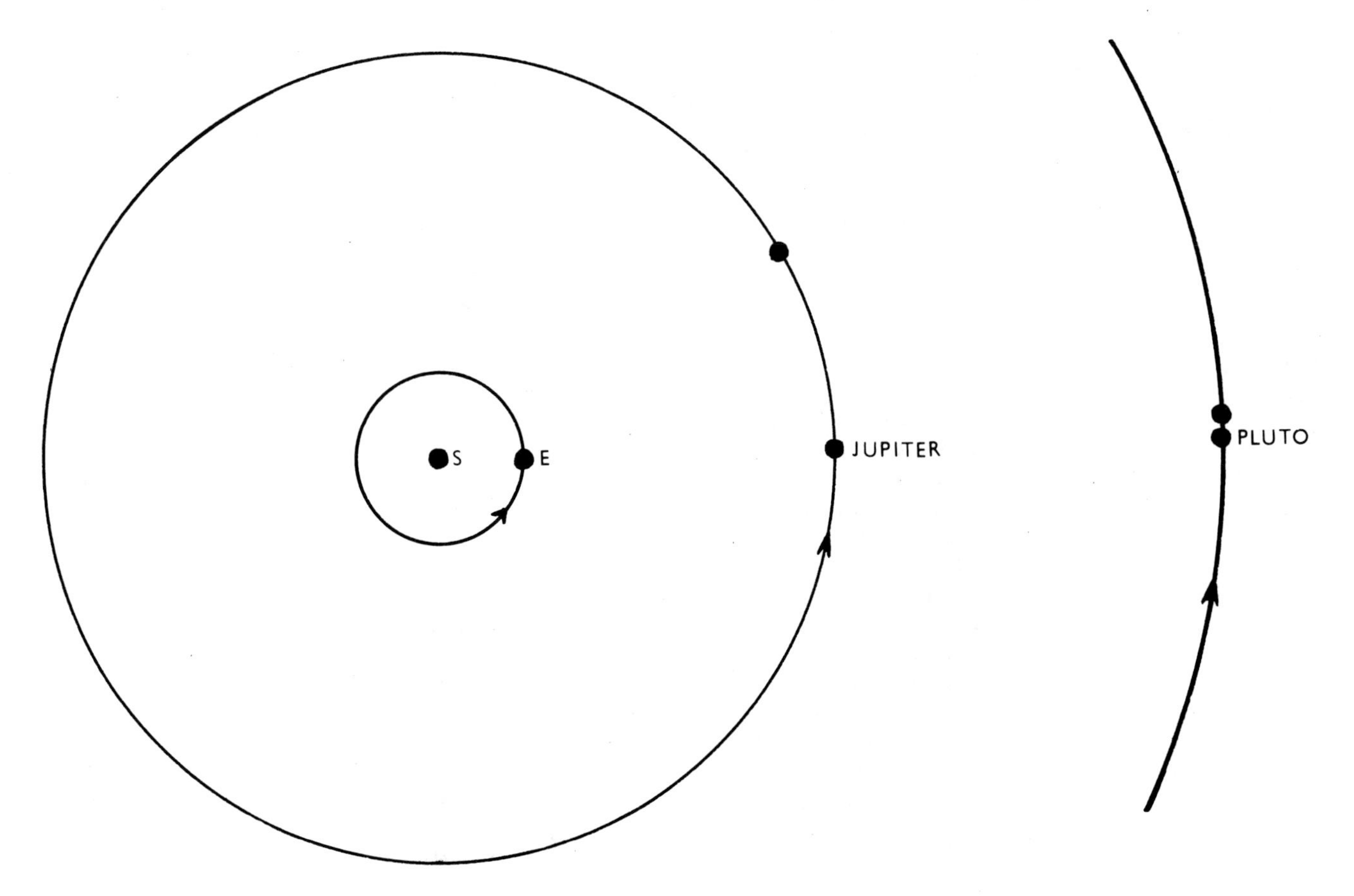

The synodic periods of Jupiter and Pluto. Within one Earth year they cover one-twelfth and one-two hundred and forty-seventh of their respective orbits.

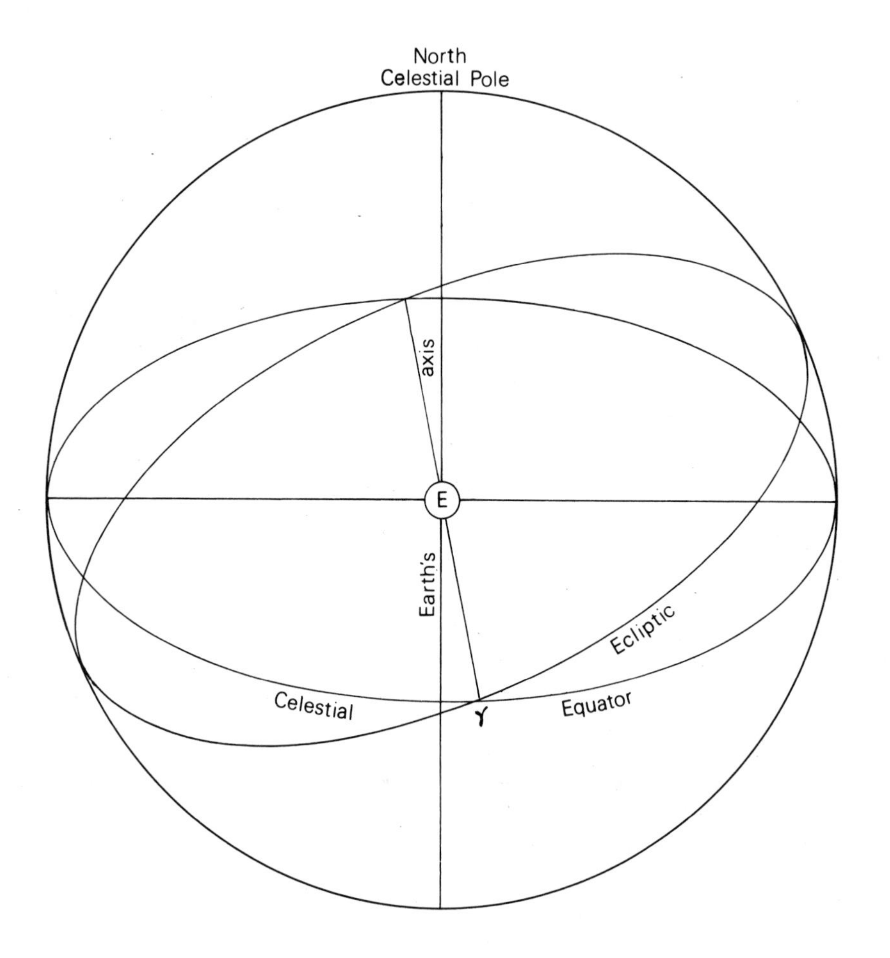

The Celestial Sphere, showing the path of the Sun (ecliptic) and the first point of Aries (γ), marking the Spring Equinox. The Sun going from south to north.

We now turn to the brightness of the planets. First, we will describe briefly the system astronomers use to measure the brightness of celestial objects, which is known as the 'magnitude scale'. When first developed, the scale started at 1.0 for the brightest stars in the sky and ended at 6.0 for the faintest stars visible to the unaided eye. Later, as the need for a more accurate scale came about, some stars were found worthy of a rating of 0.0 magnitude, while some odd stars were bright enough to require a rating even beyond this, so plus and minus signs were introduced. Thus, stars brighter than 0.0 magnitude have a minus sign, and stars fainter than 0.0 have a plus sign.

Sirius, the brightest star in the sky, has a stellar magnitude of —1.4. The faintest stars visible without optical aid are between magnitudes of +6.0 and +6.5. How visible these stars are, of course, depends both on a person's eyesight and on the sky conditions. It has been claimed that people with exceptional eyesight can on very clear nights detect stars of magnitude +7.0. This may indeed be true.

The magnitude scale

With the development of large telescopes and their incredible light grasp, stars that have magnitudes as faint as +26 have been picked up on photographic plates—thousands of times fainter than the faintest naked eye stars. To put the magnitude scale into perspective, a difference of one magnitude corresponds to an actual difference in brightness of two and a half times. A difference of 2 magnitudes corresponds to a difference of $(2.5)^2$, and so on. Thus a star of magnitude +1.0 is about 100 times brighter than a star of magnitude +6.0, that is, $(2.5)^5$ times brighter.

As we already know, five of the nine planets are easily visible without a telescope. Of the others, one is just about visible without optical aid, one needs at least a good pair of binoculars and one needs a fair-sized telescope. Saturn and Mercury, although very different from each other, usually have a magnitude similar to the brightest average stars. The brightest that Saturn can appear is —0.4. Although a large planet, it is a great distance away. Mercury can at times attain a magnitude of —1.4, similar to Sirius.

Mercury is only a small object but is quite close to the Sun and the Earth. Mars varies in brightness a great deal, simply because of its changing distance from the Earth. At times, it can outshine all the planets except Venus. When close to the Earth it can have a magnitude of —2.5 but at its greatest distance from Earth it may have a magnitude as low as +1.8. Jupiter does not vary much and at best can appear at magnitude —2.4, while never being less than —1.5 it is always a very conspicuous object.

The brightest planet

By far the brightest of all the planets is Venus. There can be few people who have not remarked on the appearance of this brilliant star in the morning or evening sky, a star which is seen long before any other in the evening and long after all the other stars have been lost in the brightening dawn sky. Even at its dimmest the magnitude of this planet is —3.4, but at greatest brilliance it can attain a magnitude of as much as —4.5. Venus is indeed the brightest object in the sky, apart from the Sun, the Moon and the occasional comet.

Venus is so bright that when conditions are right it can cast fairly distinct shadows. During late 1978 the author recalls seeing shadows cast without the slightest difficulty. Many UFO sightings can doubtless be explained with Venus. Casual and unsuspecting observers have frequently been amazed at its aspect, and most of the flying saucer sightings of late 1978 were very certainly due to the appearance of Venus. It is possible for keen-sighted people, provided they know exactly where to look, to see the planet as a tiny star-like point even in broad daylight.

Of the remaining fainter planets,

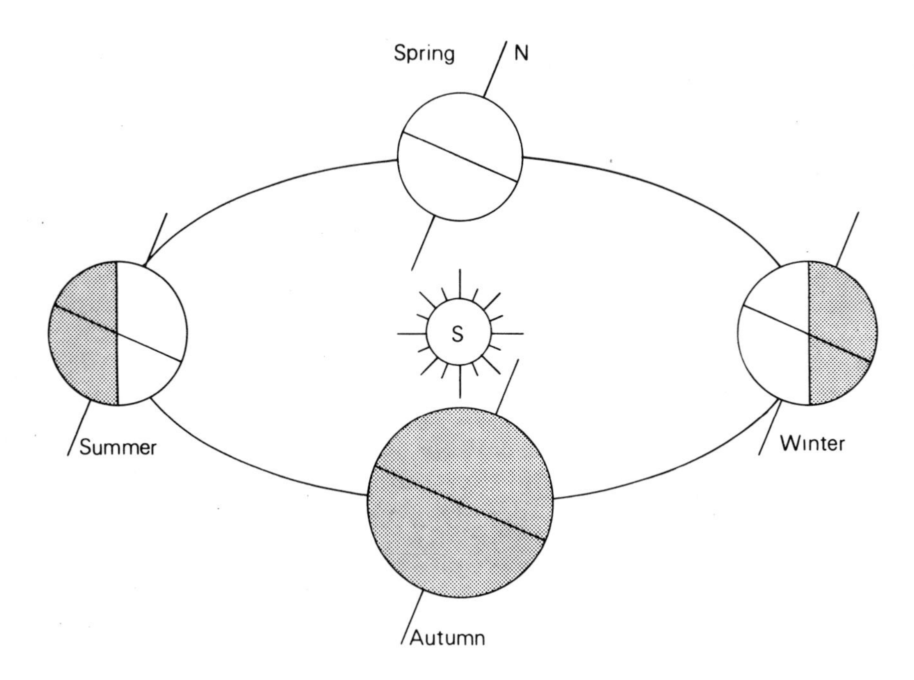

Above *As Earth orbits the Sun (S) its axial tilt remains fixed, resulting in the seasons and an apparent path of the Sun inclined to the Equator.*

only Uranus is bright enough to be seen with the naked eye. It usually shines at around magnitude +5.5 and the slightest optical aid renders it an easy object. Neptune requires some sort of optical aid to be seen. Its magnitude of +7.7 puts it well below naked-eye visibility. Binoculars will show it as a faint star, but to see it well a small telescope at least is necessary. Lastly, Pluto is so faint, at magnitude +14, that a moderate-sized telescope is needed to show it at all, and even then it is not an easy object. A telescope with a lens of 8 inches, or 200-mm aperture should just about pick it up, but a 12-inch or 300-mm aperture telescope would, at the least, be needed to show it well. Pluto is thousands of times fainter than the faintest naked-eye stars.

Reflecting power

There are many things that affect the magnitude of a given planet—its actual reflecting power, or albedo, for one thing. This, of course, depends on the material of the visible surface. Venus has a dense cloud cover which reflects light very efficiently. Mercury is a poor reflector since its surface material is very dark. There is also the consideration of the percentage of the sunlit hemisphere directed towards the Earth. With the inner planets, this percentage varies from 0% to 100%, which raises a rather important point.

The planets are not self-luminous bodies. They shine only by reflecting the light they receive from the Sun. Consequently, only half of a planet is illuminated and reflecting light. This is obviously the hemisphere directed towards the Sun, so the planet has a day, and night, side. Because of this the inner planets exhibit 'phases', like those of the Moon. When either planet is on the far side of the Sun, the whole of the sunlit hemisphere is directed towards us and the planet will appear full. As the planet moves out to greatest elongation the phase will change from full to half—intermediate phases between full and half being referred to as 'gibbous'. At greatest elongation the angle that the planet forms between the Earth and Sun is 90° and we see half the sunlit hemisphere and half the unilluminated hemisphere. Then, as the planet

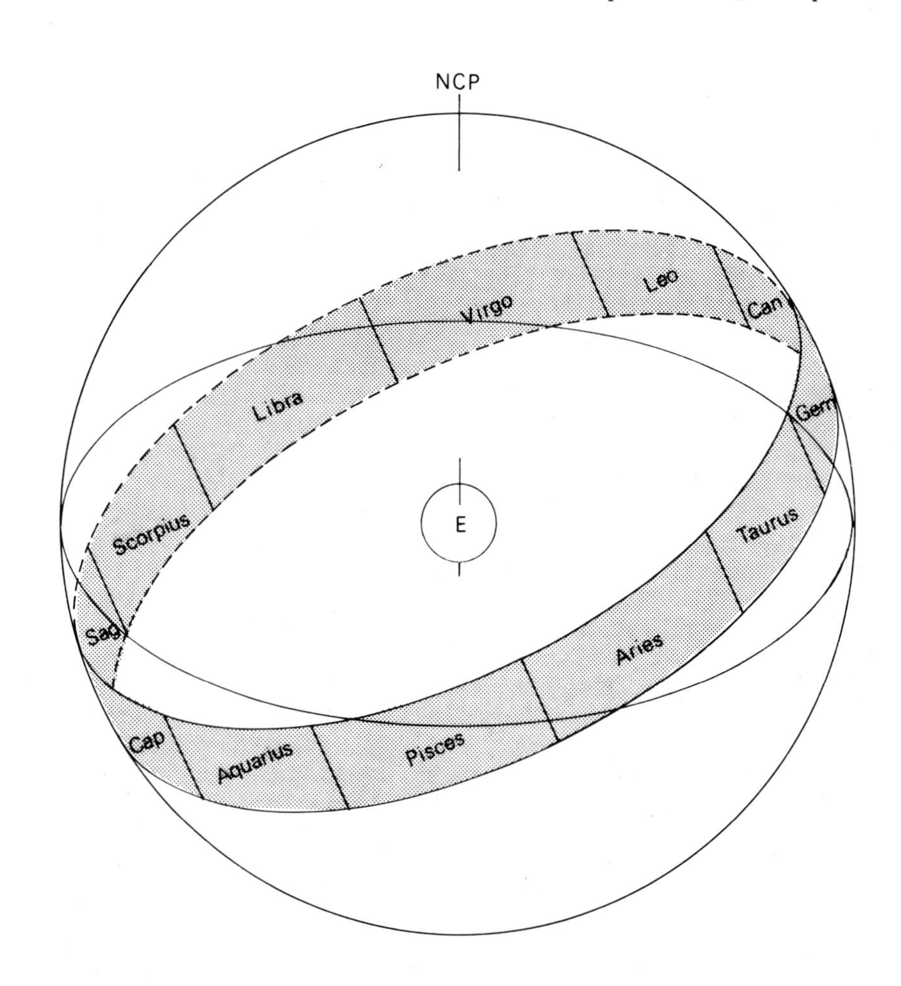

Below *The Zodiac is a band of sky about 17° wide with the Ecliptic running along its centre. All planets except Pluto will be found to move within the Zodiac.*

moves towards inferior conjunction the phase will decrease, passing through the changing crescent phases, until at inferior conjunction the whole of the night side is directed towards us. The planet is then 'new' and, in theory, invisible to us.

Apparent size

The whole situation is reversed when the planet moves from inferior conjunction to superior conjunction. It should be obvious from this that the planet's apparent size will alter, appearing much larger at inferior conjunction, when it is closer to us, than at superior conjunction. This has the effect of cancelling out, to some extent, the reduction in brightness caused by decreasing phase. In fact, Venus is at its brightest when about 27% illuminated. Between full and thick crescent the increasing brightness of Venus is governed by the increasing size of the image; from thick crescent to new the decrease in brightness is governed by the decreasing phase. For Mercury the situation is not quite the same. Its reduction in brightness, from superior conjunction to inferior conjunction, is due to the decreasing phase. Mercury's apparent increase in size is not sufficient to cancel out the decrease in magnitude. The same is true for the brightness increase from inferior conjunction to superior conjunction.

Unlike the inner planets, the outer planets do not display the full range of phases as seen from Earth. The phase effect for those planets beyond the orbit of Mars is very small indeed and does not affect their brightness noticeably. Mars, however, does exhibit an appreciable phase. The percentage of unlit hemisphere observable from Earth never exceeds 15% but it can be seen clearly with moderate optical aid. The reason for this is quite simple. When the planet is at either opposition or superior conjunction the whole of the sunlit hemisphere is directed Earthward, and the planet appears full. However, at points half-way between these two positions, known as quadrature, the Earth–Mars–Sun angle is such that we are able to see a small portion of the night side. This has some effect on the brightness of the planet, although the effect is not very great.

Magnification

Something quite different affects the brightness of Saturn. The planet is surrounded by an extensive ring system, the plane of which changes its angle with respect to the Earth so that, at times, the ring can appear quite open and, at other times, closed or edge on. (We will not go into too much detail about the changing tilt here since this will be dealt with fully in the chapter on Saturn.) The ring, when open, reflects a considerable amount of the planet's total light. When closed, we are receiving light only from the globe of the planet. The result is that with the rings open the magnitude may attain −0.4, but with the rings closed it may be as low as +1.3.

The apparent size of the planets in the sky is important when assessing how large an image observers can expect to see with certain powers of magnification. To explain this it is necessary to go into a little detail of the measurement of the sky. Two systems of measurement are used, one for the determination of longitude and one for the determination of latitude.

For longitude, measurements related to time are used. The sky is split up into 24 hours, each hour is in turn split up into minutes of time and each minute into seconds of time. These are measures of 'Right Ascension' (RA), and are exactly like the measurements of longitude on the globe. The system for latitude is measured in degrees: 360° for the whole sphere, 90° from equator to pole. Each degree is split up into 60 minutes of arc, and each minute into 60 seconds of arc. This is measurement of 'declination' and again is exactly like the measures of latitude on the globe. Provided that the correct RA and declination of a celestial object are known it can easily be located on a star map. If an astronomer has the luxury of a telescope with setting circles and the appropriate measures marked on them, and if the RA of the observer's meridian is known at the time of the observation, he can easily locate an object, once he knows its co-ordinates even if he cannot see that object without a telescope.

As far as the apparent angular size of a planet is concerned it is the second form of measurement, declination, that concerns us. We already know that degrees are split up into minutes and seconds of arc. The seconds of arc are what is used to indicate the angular size of a planet.

It may come as a surprise to learn

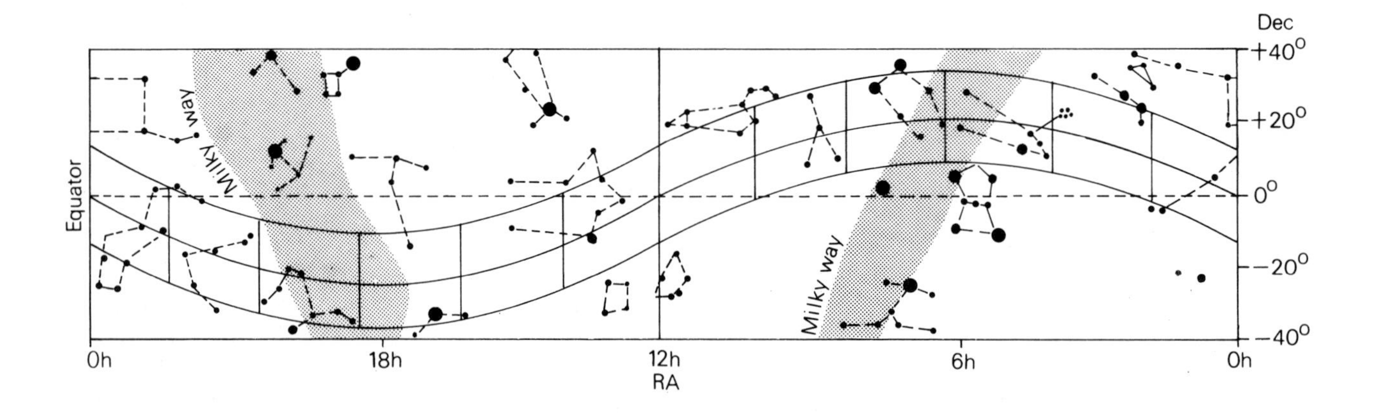

Map of the sky 40° either side of the Celestial Equator, with the Zodiac and main constellations.

that the angular diameter of the Moon is just about half a degree, or 30 minutes of arc (written 30′). If measured by a ruler held at arm's length, this would amount to only $\frac{3}{16}$ of an inch! Many people, if asked, would think the Moon much larger than this. If you are doubtful, it is worth going out and measuring it. Measured in seconds of arc, the Moon is approximately 1,800. The planets are, of course, much smaller. No set size can be given for any of them since their angular sizes vary with their changing distances from Earth. These variations have limits, however, so that mean sizes can be determined.

Angular diameter

The largest angular diameter is attained by Venus, since this planet can approach closer to the Earth than any other. Around the time of inferior conjunction it may appear just over 60 seconds of arc in diameter, written 60″ or, in fact, 1′. When at superior conjunction, it appears no more than 10″ in diameter. So, even at its largest, Venus never appears more than $\frac{1}{30}$ the diameter of the Moon. Neptune, which is a distant planet, does not appear to alter in size too much since the changing position of Earth has little effect on so remote an object. Its diameter is around 2.5″. The following are mean opposition angular diameters for the outer planets and mean inferior conjunction angular diameters for the inner planets.

Mercury	10″
Venus	60″
Mars	17″
Jupiter	46″
Saturn	19″
Uranus	4″
Neptune	2.5″
Pluto	0.2″

The value for Jupiter refers to its equatorial diameter; that for Saturn to the equatorial diameter of the globe and not to the rings.

The size that a planet will appear through a telescope depends upon the magnification used. A rough idea can be gained from the fact that a magnification of 72 times (×72) would give an image for Mars of about the same size as the Moon seen with the unaided eye (this, of course, being when Mars is situated at one of its closer approaches to the Earth). As for Jupiter, a magnification of only ×40 would give a similarly sized image.

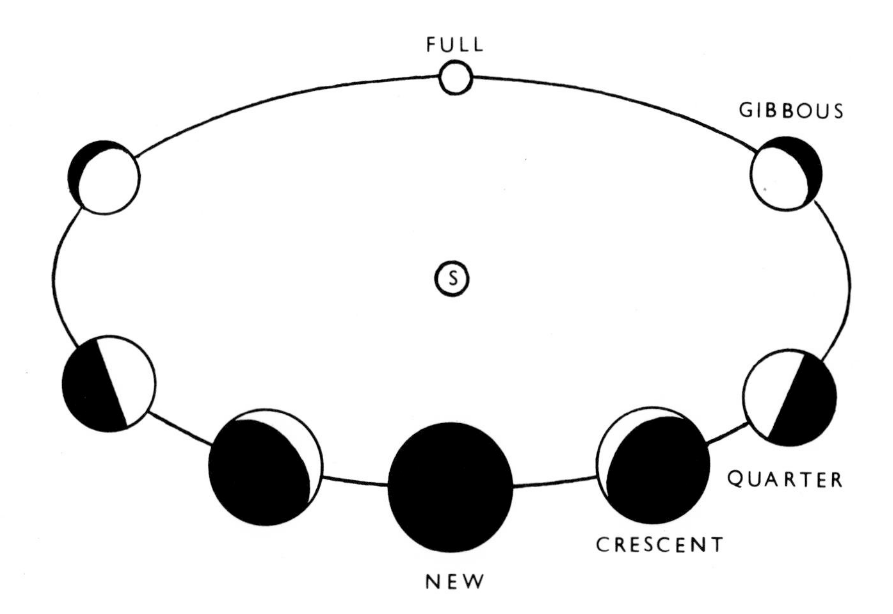

Appearance of phases for the inner planets as seen from Earth.

It is just because the planets do have a certain angular size that they do not appear to twinkle like the stars. The stars are so remote that they have little or no angular size and are mere points of light. Their light is severely affected by the atmosphere of the Earth, and the slightest turbulence makes them appear to flicker and even to change colour rather rapidly. The effect becomes more obvious the lower an object is in our sky. This is where the dense layer of atmosphere has a considerable effect and, as mentioned, destroys the clarity of the telescopic image. Many readers will have noticed on winter evenings a very bright bluish-white star, low down in the southern sky, which almost appears to flash in and out and change rapidly through many colours of the rainbow. This is Sirius, the brightest of all the stars. Because of its poor altitude, as seen from the Northern Hemisphere, Sirius is badly affected by atmospheric turbulence, and its brilliance makes this more noticeable. The flashing effect is known as scintillation.

Colours

Although not readily visible as discs to the unaided eye, the planets do have tiny measurable images and are less obviously affected by our atmosphere, so that they appear to shine with a steady light. This is a good way of distinguishing them from the stars without the use of a telescope. The colours of the planets are also, in some cases, quite striking. Mercury usually has a pinkish colour, although this may be due to its low altitude and to the reddening effect of the denser layers of our atmosphere and may not be the true colour of the planet. Indeed, for this planet to be seen with the naked eye it must always be close to the horizon. Venus appears as a brilliant white, or even silver, but its sheer brightness is enough to make it instantly recognizable. Mars has the strongest colour of all the planets and truly earns its name of the 'red planet'. As a fiery orange object it is easily recognizable. Jupiter is bright yellowish white and Saturn seems to have a pale yellowish orange lustre. The fainter planets, of course, cannot easily be seen without a telescope but, when so viewed, the two brighter ones display some coloration. Uranus has quite a distinct bluish green colour, while Neptune appears blue-white. With a little experience it becomes possible not only to recognize the planets with the unaided eye but even to identify them individually.

Information on much of what has been discussed may be found in many astronomical handbooks or almanacs. The *Nautical Almanac*, *The Astronomical Ephemeris* and the *Handbook of the British Astronomical Association* are among the most useful. In addition, there are several publications that give all the astronomical events of interest for the coming year. From these one can obtain information on the apparent angular diameters of the planets for any given time. Phase percentages are also given at regular intervals and magnitudes, opposition and conjunction dates and positions of the planets throughout the year. You will find, in addition, a great deal of other information that is invaluable to the beginner and serious observer alike. Literature can be useful.

Disappointment

When armed with all the relevant information and a fairly good telescope, the time is right for the would-be observer to go out and have a look for himself. But if he is not careful he will be bitterly disappointed and become totally disillusioned. Many is the time that an observer has taken a quick look at the planets and, because all the details he may have seen on photographs or drawings made by experienced astronomers and reproduced in various books are not immediately resolved to him, thinks that something is wrong with either his equipment or his eyesight. Usually nothing could be further from the truth. Several things may be responsible, as you will have already gathered from the earlier parts of this chapter. First and foremost is the simple fact that a quick glance is certainly not enough. Even the most experienced observer has to spend some considerable time at the eye-piece of the telescope familiarizing himself with any detail present. Only when he is entirely satisfied that he has seen virtually all there is to see can he start recording the details by making notes or a sketch. Even while in the process of recording the details, he may see features he has not noticed before.

Another possible cause of disappointment on one's first viewing is the condition of the atmosphere. Remember the twinkling of the stars? This, we recall, signifies turbulence in the Earth's atmosphere which may be worse on some occasions than on others. If you go out to observe and you see that the stars are twinkling very noticeably, then it is probable that it will not be worth your while attempting planetary observation. The sky may look perfectly clear, with the stars standing out like jewels, but the view of a planetary image through a telescope will tell a different story. The planets may not twinkle to the unaided eye but their magnified image will appear to shimmer, and even dance, like something seen through a heat haze, as the turbulent air currents pass across the field of view. This is called 'bad seeing' or 'boiling'. You will only become frustrated by continuing to look at such an image and the best advice is to give up and try again later. Do not force yourself to detect any detail since this is likely to be spurious. Later in the night the air may well have settled down and the view will be much improved. Remember, no matter how good a telescope you have, it will not be able to cope with poor seeing conditions that you will meet.

Appearance of phases for Mars.

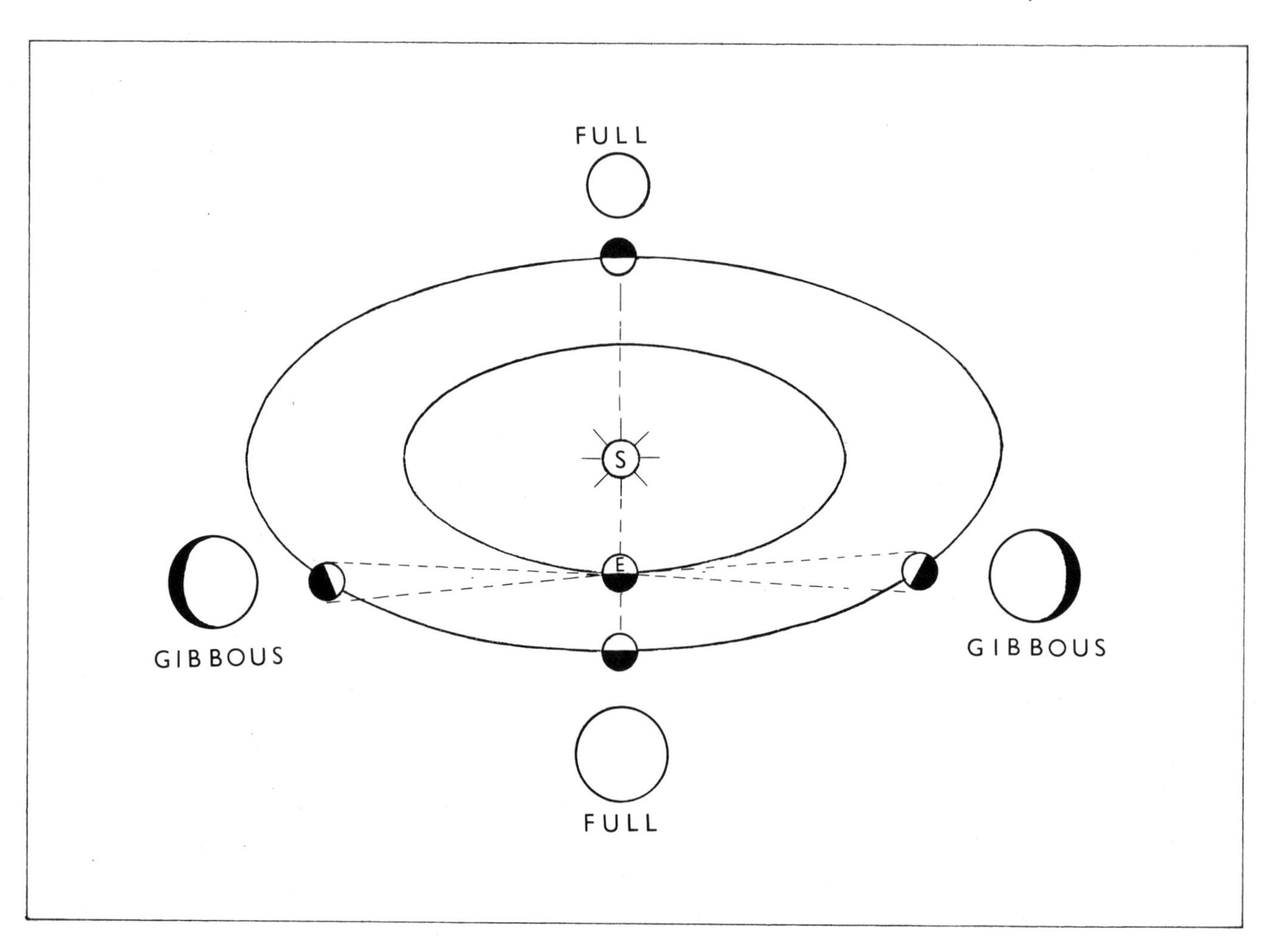

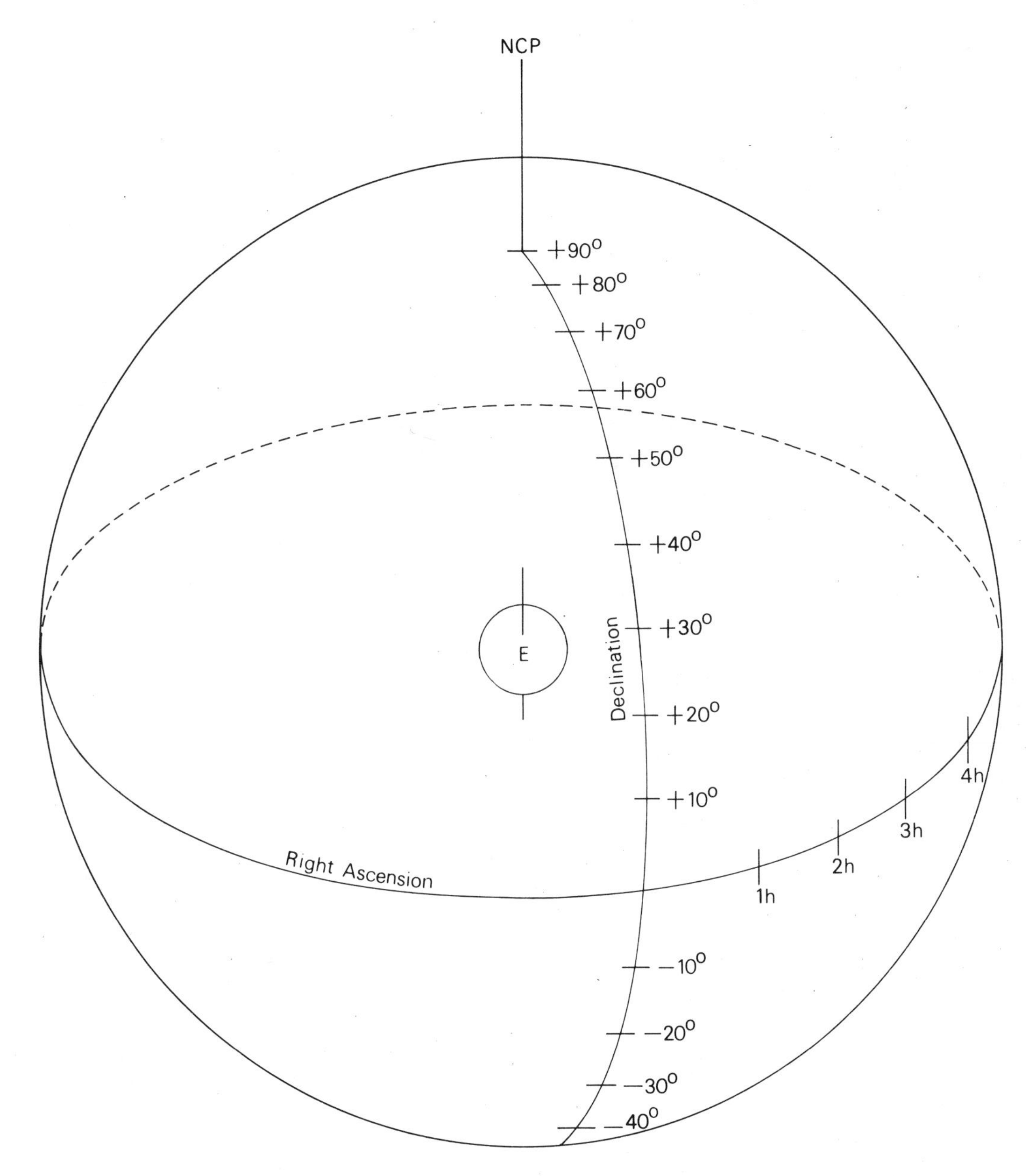

Right Ascension and Declination on the Celestial Sphere. (See page 20 for these coordinates as they appear on a map of the sky.

Many people are surprised to learn that the best views of planetary images are obtained when the sky is a little misty, or with conditions bordering on a slight fog. At these times the air is very steady and the planetary image will as a result appear sharp and still. There will be some loss in brightness, which is not ideal for faint objects that may be surrounding a planet, but this will be compensated for by the advantage of a steady image. Occasionally one will experience times when the air is transparent and still, enabling you to get a bright steady image but these are rare and should be put to the best possible use. These conditions can almost be classed as 'perfect seeing'. In the author's experience, the number of such nights over the last five years can almost be counted on the fingers of one hand.

Choice of planet

Finally, the observer's choice of planet may not always be a wise one, which may again lead to disappointment. For instance, one cannot expect to see a great amount of detail on the planet Venus, for the simple reason that it is not there. Mars is not always an ideal choice, either. It spends a lot of time at great distances from the Earth, which often renders the disc of what is, after all, a small planet, too tiny for detailed work. Some experience is needed for this planet and only when it is viewed at favourable times can useful observations be carried out. Jupiter is possibly the best choice for the new observer with perhaps only a fairly small telescope, since this planet nearly always presents a wealth of quite prominent detail.

It is of major importance to the success of any observer that he takes note of the various situations concerning the favourability of a planet and, provided he picks the right time, the right planet and the right conditions of seeing, he will not be deterred from carrying out serious work. The author can assure readers that there is enormous satisfaction to be gained from taking the trouble to do this.

Mercury

Unfortunately Mercury is never particularly well placed for observation, because of its closeness to the Sun. Even at its most favourable elongations it never attains a distance of more than 27° 45′ from the Sun and, since the orbit of Mercury is rather eccentric, this figure is often much less, the minimum elongation distance being 17° 50′. As a result, it never sets much more than two hours after the Sun or rises more than two hours before it. Mercury is never seen in a dark sky and is usually best seen with the unaided eye at twilight. Therefore, despite the fact that it is one of the five bright planets, it is often difficult to spot and may easily go unnoticed. Even at those times when it is seen without optical aid the view through a telescope is pretty hopeless owing to the planet's low altitude and the resulting bad atmospheric conditions.

The actual mean distance of Mercury from the Sun is 57.9 million kilometres although, because of the eccentricity of its orbit, it can at times approach to within approximately 45.6 million kilometres and at others be as far as 69.6 million kilometres. This means that at inferior conjunction it may approach to within 76 million kilometres of the Earth. Since it is the smallest of the major planets, with a diameter of only 4,880 kilometres, its angular diameter from Earth, never exceeds 12″. Detailed study is thus, at best, far from easy and a fairly moderate-sized telescope is needed before serious work can be undertaken. Furthermore, this value of 12″ (page 20) for the maximum angular diameter applies only to the time when Mercury is at aphelion (its greatest distance from the Sun) and at inferior conjunction, when theoretically invisible to us. When near superior conjunction its disc is hopelessly small at 5″. So the best time to attempt observation is during the few weeks around greatest elongation.

At each elongation Mercury can be followed for about five or six weeks and the phase during this period will range from a fairly thick crescent to about three-quarters full. The angular diameter will range from between 6″ and 9″ and a magnification of ×200 to ×300 will therefore give an image size comparable to the unmagnified image of the Moon. At either conjunction it is much too close to the Sun for observation by amateurs.

Adequate results

A telescope with at least 200-mm aperture is required to give the planet a usefully sized and clear image, although it has been claimed that a telescope of only 150 mm will give adequate results. It should be said in all fairness, however, that for serious study apertures in excess of 300 mm are essential. Mercury tends to be neglected by most amateurs for this reason.

Because of its short 88-day sidereal period and average 115.9-day synodic period, elongations are frequent, if somewhat short, and give plenty of opportunities for spotting the planet during the year. The average number of elongations taking place in a year is six—three morning, or western, and three evening, or eastern. For observers in Britain, the most favourable evening elongations occur between March and June and the most favourable morning elongations between August and October.

Because of the very difficult conditions of seeing encountered when trying to observe this planet near the horizon, serious observers prefer to locate the planet during the daylight hours, when it may be found quite high in the sky, well away from horizon haze. Contrast at these times will be very poor because of the brightness of the sky. The superior conditions of seeing, however, make this inconvenience a necessary evil. Possibly the best possible time to locate Mercury is a couple of hours either before sunset or after sunrise, when the air tends to be much steadier than at mid-day. Contrast is also much better as the Sun is closer to the horizon, and the sky is less brilliant. Noon is by no means the best time to view; the heat from the Sun can have a considerable effect on seeing conditions.

The method of finding Mercury during the day is not too difficult, provided the telescope used is on an equatorial mount, that is, fitted with setting circles in both RA and Declination. One has to be very careful as it is the Sun that is used as the guide for locating the planet, and the Sun should never be viewed directly through a telescope. Point the tele-

Lining a telescope on the Sun. It is safer not to remove the lens cap. Once the telescope is directed to the Sun, any position in the sky can be found by moving the telescope a measured amount provided the coordinates of the Sun are known. This makes the detection of the planets possible during the day.

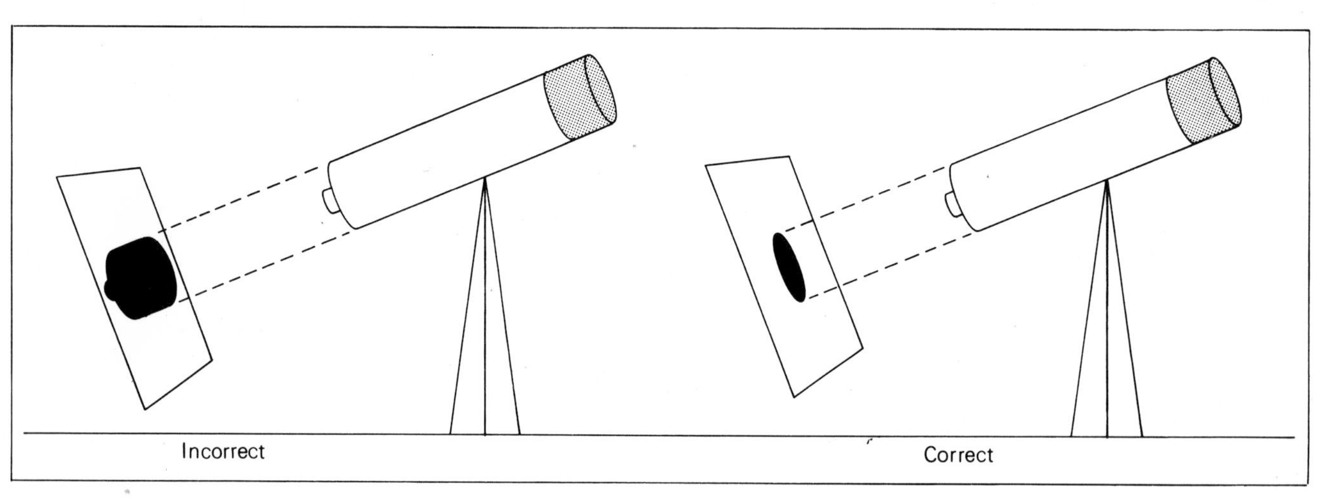

scope at the Sun by looking at the shadow the telescope tube casts on the ground. When this is circular, the telescope is lined up with the Sun. To make sure, project the Sun's image through the eye-piece on to a piece of white paper or card. The observer must know beforehand the relative positions of the Sun and Mercury. This information can be obtained from astronomical handbooks.

Suppose that the planet is following the Sun by one hour of RA, then move the telescope eastward by this amount without elevating or depressing it. The telescope will now be pointing to a position in the sky of the same RA as Mercury, but still of the same declination as the Sun. Mercury may well be higher or lower, that is, N or S of this position. The difference in Declination is obtained from the handbook. If, for example, it is 10° north of the Sun the telescope must be elevated by this amount. With luck, and provided the telescope is accurate, Mercury will now be in the field of view. To assist here, a fairly low power should be employed, giving a reasonably wide field of view. If the power is too low, however, the image will be so small that there is a possibility of missing it. If the planet is not in the field then a little very careful sweeping will probably locate it.

During daylight hours the contrast of the image against the bright sky is very low, and sweeping too quickly will result in it being overlooked. You will also need a crystal-clear deep blue sky since the slightest haze will render the pale image invisible. It is most important to have the telescope focused correctly. This is often forgotten and it is sometimes quite difficult to find something during the day that is distant enough to give the same eye-piece position required for focusing on Mercury. The best thing to do, of course, is to select the eye-piece you will be using to locate Mercury and, on a clear night, focus on a star and mark the eye-piece drawtube for future reference. If however, you have omitted to do this then the telescope should be focused on a very high aeroplane. This will not be far off the focus for Mercury. Moreover, if the Moon happens to be in the sky at the time then this will give you something else to focus on. Focus is most important since, even if slightly out, it will result in the planet being missed.

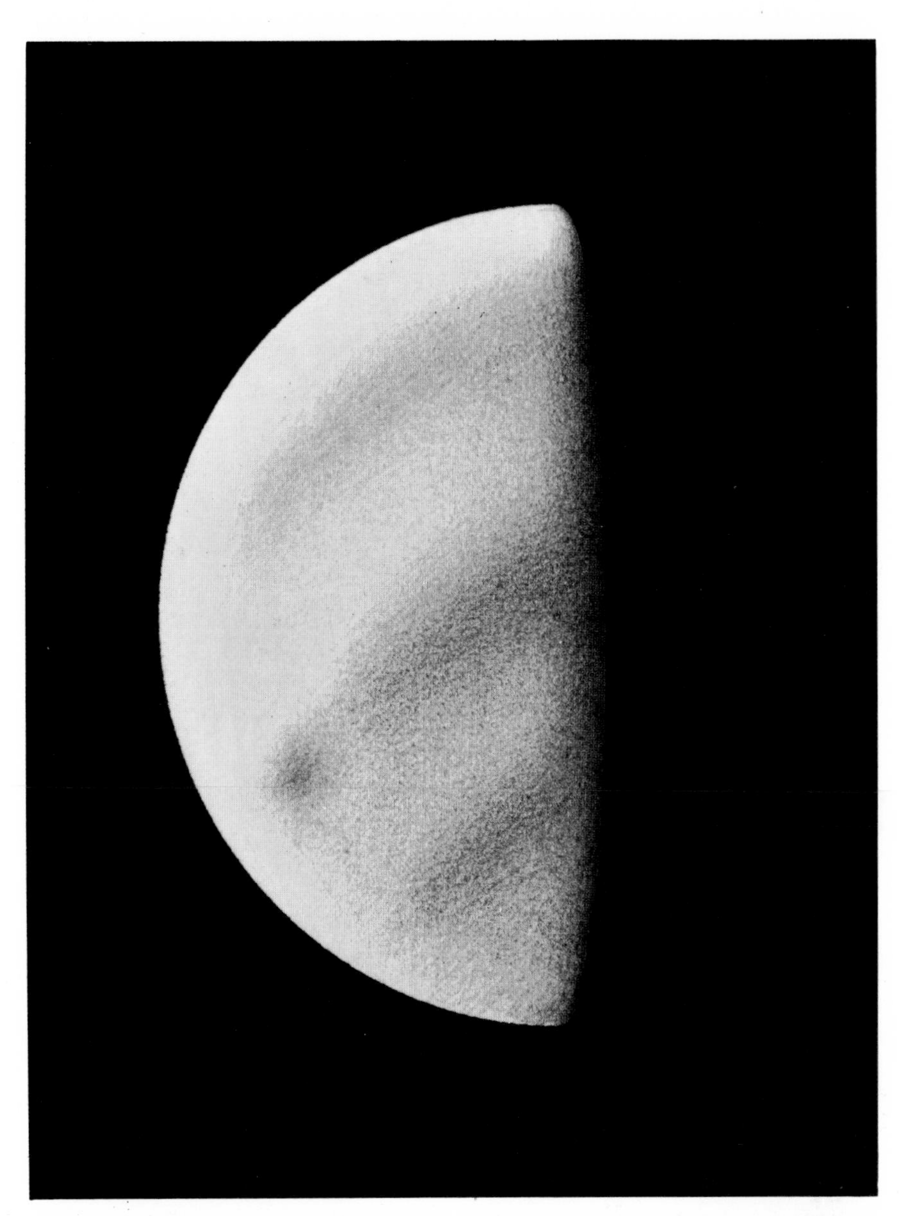

22.IV.76 19.05 254mm Refl ×300 Mercury's faint markings.

Locating the planet

For telescopes that are not fitted with setting circles the easiest way of locating the planet is again to direct the telescope towards the Sun in the manner described earlier and then to wait for a period of time corresponding to the difference in RA. It would be a good idea to cover the lenses of the telescope while waiting for the time to pass, because as the sun moves out of alignment with the telescope optics its image will for a time be projected on to the inside of the telescope tube wall and may cause damage by burning it. It can be very distressing to see your valuable telescope going up in flames. Covering the lenses will prevent this and will also stop anyone from taking a glimpse through the telescope while the Sun is still in the field.

Once the required time has elapsed, alter the elevation of the telescope by the difference in declination, either north or south. This difference can be estimated by considering the diameter of the solar disc which is just about $\frac{1}{2}°$. If Mercury were almost 5° north of the Sun then you would elevate the telescope by an amount corresponding to 10 solar diameters. This second method will only work when Mercury is east of, or following, the Sun and should not be used when Mercury is at western elongations.

An imaginary view of the surface of Mercury. Its scorched and rugged appearance is similar to that of the Moon. Even with the Sun shining the sky is black owing to the lack of atmosphere.

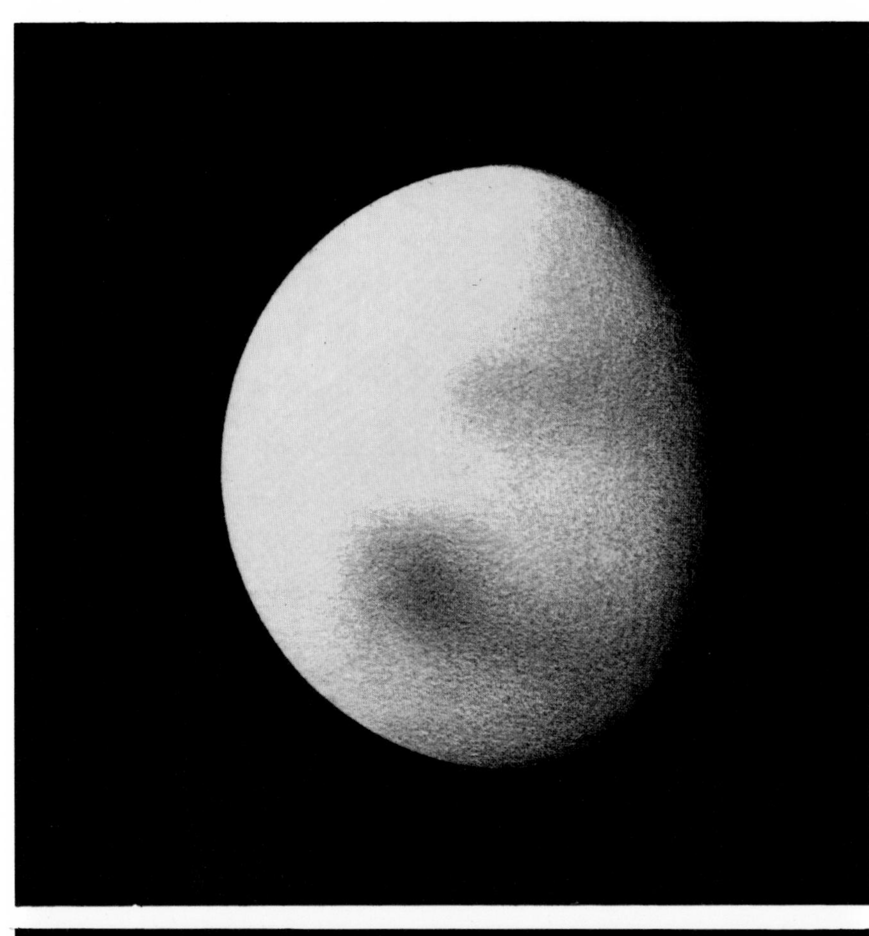

A third, though less practical, way of locating the planet during the day is to direct the telescope towards a star on the night before the intended observation. This star should, of course, have the same declination as the planet. If the difference in RA between the star and the planet is known, it will be an easy matter to work out at what time on the following day Mercury will have the same position as that held by the star. For example, a difference of 14h in RA will place Mercury in the same position 14 hours later. The greatest problem here is the high probability of a change in the weather. Unless there is shelter for the telescope, it is unwise to leave it unprotected for long periods. I do not know of many people that would give odds on the possibility of a 14-hour rain-free period in this country.

The best time

The best time at which to locate the planet is shortly after superior conjunction. For then, although the disc is tiny, its surface brightness is fairly high and it stands out quite well even in a bright sky. Often the brilliant Venus will occupy the same area of sky and will act as an excellent guide to Mercury's location. Perhaps the easiest method of all is to find Mercury when it is a morning star, low down on the eastern horizon in a fairly dark sky and an obvious naked eye object. Point the telescope towards it and retain it in the field for a couple of hours as it climbs higher into the sky. It may be followed in this way until long after sunrise.

Finding the planet leads on to other, more difficult things to look for and their difficulty adds to their im-

Above left *28.III.77 16.00 UT 419mm Refl ×248* Left *15.IV.77 19.20 UT 254mm Refl ×230* Above right *12.III.78 17.50 UT 419mm Refl ×372* Right *29.III.78 17.50 UT 419mm Refl ×372 Because of the relationship between Mercury's rotation and the position of the Earth in its orbit the same markings are seen under similar conditions. This led to thinking that Mercury's period of rotation was exactly the same as its orbital period, as shown here.*

portance. Firstly, an accurate estimation of the percentage of the illuminated portion, or phase, should be attempted. Ordinary visual estimates will suffice. The easiest way is to sketch the terminator, that is, the division between the day and night hemispheres, as accurately as possible, on to a circle 50 mm (2 inches) in diameter, and measure the proportions of the illuminated and unilluminated parts across the full diameter of the disc. With practice one can become quite accurate. If a large number of observers are carrying out this sort of work, their estimates can be collected together at the end of the elongation and averaged out. The more observations, the better the average, as any discrepancies will be smoothed out.

You will find the 'observed phase' usually differs quite noticeably from the theoretical. Many reasons for this, the so-called 'Schröter effect', have been put forward. The falling off of brightness towards the terminator no doubt plays a vital part in making the phase appear less than it should, particularly if the contrast is poor. The brighter the sky the smaller the percentage, since one loses sight of the fainter regions. The phase always appears greater when the image is bright. Seeing conditions also play their part. The deviation from the theoretical phase is most noticeable at the time of dichotomy, that is, the time when the planet should appear exactly half. On average, it seems that observed dichotomy occurs about one day early at evening elongations and one day late at morning elongations if compared with theoretical dates, which does tend to suggest that a fall-

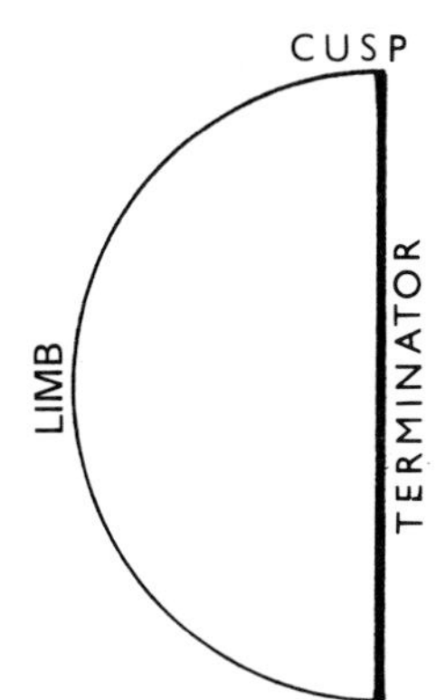

Names for the various parts of a planet's disc. Dichotomy refers to the exact time of half phase.

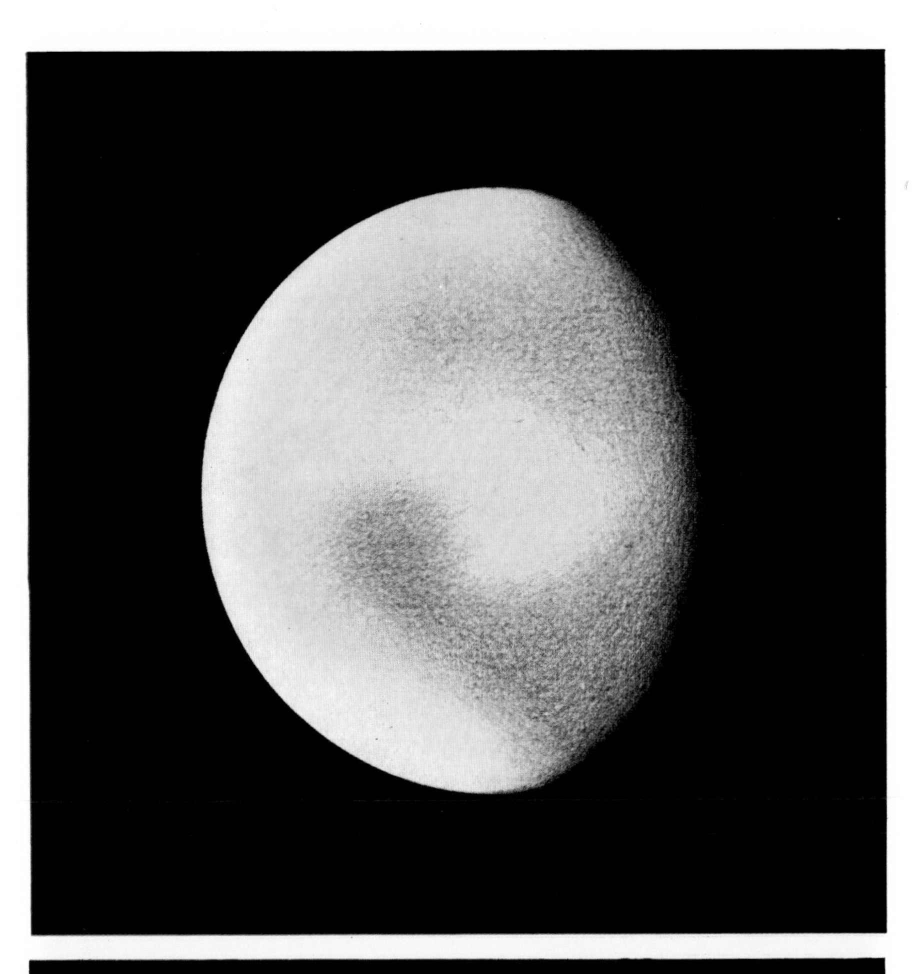

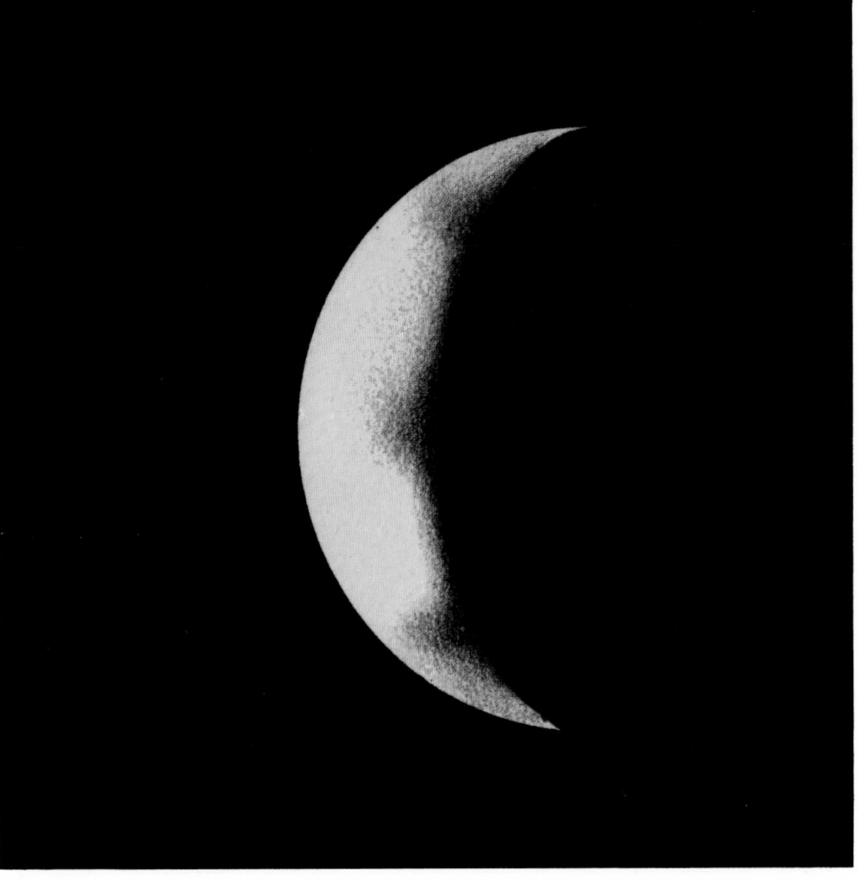

off of brightness near the terminator is the cause.

Various other phenomena have been associated with the planet Mercury. The cusps, for instance, on the crescent, do not always look the same. One may appear sharp, the other blunt. A similar effect can occasionally be seen when the phase is half or just over. This may be due to a particularly dark marking in the region of one cusp making it appear blunt, while a bright area near the other cusp may give it a sharper appearance. In the past, it has even been suggested that bright dust storms or clouds near the cusps could have something to do with this, but since we now know that the planet has only a very tenuous atmosphere this possibility can be ruled out.

Strange phenomenon

There have been occasions when discrepancies have been noted in the terminator, these taking the form of either dents or humps. These cannot be true surface irregularities since there is no mountain high enough, or crater deep enough, to give rise to discrepancies of a sufficient size to be observable from the Earth. Some observers of the past, notably J. Schröter, have on occasions reported small bright points of light appearing, detached slightly from the cusp or terminator. This was attributed to a mountain, the summit of which was catching the rays of the sun before the terminator had reached the vicinity of the mountain. Since such a mountain would have to be at least 80 kilometres high this does not seem likely.

The apparent variability of observed markings has caused some mystery. Dust storms and atmospheric veils were used to explain these but they probably result from the real difficulties of observing any marking on the planet. It may be that solar radiation, which will be considerable at the position of Mercury, is having some effect on the material that makes up the planet's surface. This is something that only continued observation will resolve. A brightening of the 'limb' (the true edge of the planet's disc) is often observed, as is shown on some of the illustrations presented here. Most of these are phenomena that could easily be explained if the planet had a considerable atmosphere, but this is unfortunately not so. Some may be attributed to pure optical effects, others to varying angles of illumination.

With Mercury, we are seeing the true surface and it has been possible to draw up maps showing the various permanent markings. First to do this with any success was the Italian astronomer G. Schiaparelli. His observations, made towards the end of the last century, were carried out during the day. An even better map was later compiled by the Greek astronomer E. M. Antoniadi, who used the giant telescope at the Meudon observatory in France. Again his observations were made during the day and some of his drawings of Mercury are among the finest ever produced. Both of these observers drew up their maps on the assumption that Mercury had what is called a 'captured rotation', that is, one which is exactly as long as its sidereal period, resulting in the planet keeping one hemisphere directed towards the Sun all the time. Later, in the 1950s, improvements to these observations were made by A. Dollfus and H. Camichel, who used the telescope of the Pic-Du-Midi, situated high in the Pyrenees, to observe the planet under superb atmospheric conditions — though again their map assumed that Mercury had an 88-day period of rotation.

Radar observations

Radar observations during the mid-1960s, however, showed the planet to have a rotation period close to 59 days, which meant that astronomers needed to take a fresh look at their observations. Interestingly, this new rotation period has not shown the earlier observations to be completely wrong—in fact, quite the opposite. It so happens that Mercury's rotation period is two-thirds of its sidereal period, which ties in rather remarkably with its synodic period. This means that when the circumstances for observation are the same, the part of the surface directed earthward will also be the same. So it is perfectly understandable that the earlier observations were misinterpreted, particularly in view of all the difficulties.

Armed with information sent back by radar, a new look was taken at the earlier observations and a rotation of 58.6 days was derived from them. A new 'albedo' map was drawn up by C. Chapman of America, followed by the highly successful Mariner 10 space probe, which made three close approaches to the planet during 1974–75, sending back pictures of incredible clarity and increasing our knowledge of the surface far beyond our expectations. Astronomers have never been really happy with their knowledge of Mercury. Now, Mariner 10 has furnished us with detailed views that could never be matched by Earth-based telescopes.

The planet's surface

The surface of the planet Mercury resembles most strongly that of the Moon, being covered by craters and dark areas. It was suspected from telescopic observation that this would be the case since the albedo of Mercury is almost the same as the Moon's. Also, the radar observations, mentioned earlier, did indicate a very rough surface.

The space probe certainly dealt a blow to the usefulness of amateur work, and it is often said that Mercury is no longer worth looking at. But do not be put off. Spacecraft can only remain in the vicinity of the planet for a limited time and it is still possible for Earth-based astronomers to find things out by continual monitoring.

Recording details on the planet is not easy, for the features are elusive. Experience of observing the planet on many occasions with 25-mm, 419-mm and 450-mm reflecting telescopes show that the larger instruments are a great advantage. In good conditions, the markings can be clearer than the faint shadings of Venus and fall in prominence between the features of Venus and those of Mars. Conditions are often not good though, hence the difficulty. One observation made by the author in 1969 with the 450-mm reflector showed markings that in some ways match those of accepted maps but there were also differences that were difficult to account for. The observed markings actually agreed with those of an observer using a much larger telescope during a different elongation, although different portions of the globe were on show on each occasion. This example serves to illustrate the difficulties.

Spring is the best time. A series of observations over consecutive spring

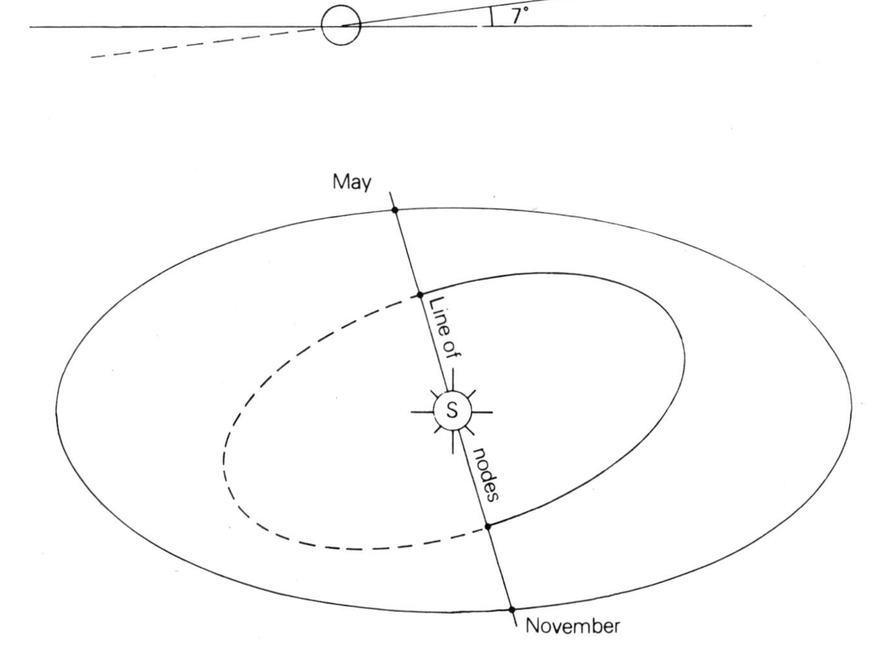

Diagram showing how transits of Mercury occur.

evening elongations may often be carried out in ideal conditions. The markings observed change very little from day to day throughout the period of the elongation, which is to be expected. Also, there is good agreement between features observed from year to year. This explains how the original assumption of a captured rotation came about. Generally, the most obvious features have been found on the northern half of the disc and some slight variation in the intensity of the markings has been detected. This remains a mystery.

Varying colour

On more than one occasion, certain dark markings have given the appearance of dents in the terminator. The southern cusp in general has been the darker while the northern has appeared quite bright. The phase has been consistently less than that predicted by between 2% and 3%. Four illustrations are presented here, showing the appearance of the planet on two separate spring evening elongations. Patience is needed in the recording of these details and, since good observations do not come easy, each one must count. Things need not be quite as bad, however, as they are often made out to be. If observers are prepared to try hard enough, then they will surely meet with success.

The colour of Mercury is remarkable. When observed with larger instruments during the day the planet tends to look pinkish or yellowish pink, but when seen in a darker sky with the image brighter, it is more yellow.

The author recollects a very fortunate occasion in 1976. It happened that on one evening the crescent Moon was in a similar part of the sky to Mercury and so placed that, by keeping both eyes open, it was possible to see the planet with one eye at the eye-piece of the telescope and see the Moon with the other eye over the top of the telescope tube. In this way the image of the planet could even be brought close to the image of the Moon. The magnification used at the time was such that the two appeared to be of similar size with both objects having an identical crescent phase. Since they were in the same part of the sky they were both affected to the same extent by the sky brightness. What stood out was their similarity of colour. Both appeared pale yellowish white but with the slightest hint of pink in Mercury's image. It was a very lucky set of circumstances. There has been much disagreement in the past about the planet's apparent colour and no doubt atmospheric conditions and altitude have been responsible. There can surely be no better way to judge the colour than by comparing it with another celestial body, situated so that both are affected in the same way.

One type of observation is certainly not difficult to make. You may recall from the last chapter that the greatest apparent angular diameter of Mercury is attained at its inferior conjunction, when the planet is new and theoretically invisible and the whole of the night hemisphere is directed towards us. This is by and large quite correct, but there are occasions when the planet can be seen at inferior conjunction—when the planet actually passes in transit across the solar disc.

The orbit of Mercury is inclined to the plane of the ecliptic by about 7°. The points where its orbit crosses the ecliptic are referred to as the 'nodes'. Having a short orbital period, the planet often passes through inferior conjunction but, because of this inclination, it more often than not passes either north or south of the Sun as seen from the Earth, missing the actual disc of the Sun. However, at the times when it is situated at, or near, the nodes, a transit will occur.

The position of the ascending node of Mercury corresponds to the position that the Earth occupies in its orbit on about November 9. That of the descending node corresponds to our position on May 7. So transits can only occur at inferior conjunctions around these dates. During November the Earth is slightly closer to the Sun than it is in May. Thus, the apparent diameter of the Sun is larger during November. Consequently, transits can occur as much as five days either side of November 9. In May, however, transits will only occur if inferior conjunction happens in a period three days before or after May 7. For this reason, November transits are much more common than those in May. The minimum time between any two transits is at present three years, the most recent example being the May 1970 and November 1973 events. The maximum period between successive transits is thirteen years, which is the period we are now in, and the next transit will not occur until November 13 1986.

When in transit Mercury appears as a small black spot, silhouetted against the brilliant disc of the Sun. This spot will appear slightly larger during the May transits since the planet is then closer to Earth. Even so, its angular diameter is still only 13″. The duration of a transit may vary considerably, depending on several factors. If the planet passes centrally across the disc of the Sun then the transit may last up to 8½ hours. This was almost the case in 1973. However, if the planet merely grazes the Sun's limb, as it will in 1999, it may only last for a few minutes. The length of time that the planet usually takes to emerge fully on to the Sun's disc is roughly between three and five minutes (this may again vary considerably depending on the position of the planet's track across the Sun).

Observing transits

Once again, observation should not, of course, be by direct viewing, but by projection through the telescope on to white paper or card, as explained earlier in the method for finding Mercury in daylight. The solar image is so bright that it may be considerably enlarged in this way and still remain clear. The projection must be done through the eye-piece, and some way of shading the paper found in order to show the image up. A large piece of cardboard with a hole cut into it, slipped on to the telescope tube, will be adequate if a refractor is used. If you use a reflector, anything to keep stray light away from the paper will do, although the shades must not, of course, interfere with the projected image. Large apertures are unnecessary and impractical here; satisfactory observations can be obtained with only a 50-mm refractor.

On first looking at the Sun's image the planet might be difficult to pick out. If the Sun is going through a particularly active sunspot period, which it tends to do at regular intervals of eleven years, then numerous sunspots will be found on the Sun's disc that will tend to confuse things. (Sunspots are those areas of the Sun's surface that are a couple of thousand degrees cooler than the general surface of the Sun.) With a little practice, however, the planet becomes easy to pick out because of its roundness and blackness. It will appear sharper in outline than the sunspots, which are surrounded by less intense areas, or penumbra, that tend to give the spots a somewhat diffuse appearance. Apart from this, if you maintain a watch for some time you will see that Mercury is moving quickly relative to the spots.

Another method of observation concerning transits, and one which is not recommended unless the observer is of considerable experience, is by direct viewing with the use of a solar filter. The filter is placed over the objective of the telescope in order to reduce the light of the Sun to a safe level. Special filters may be purchased for this purpose, but their use by the average observer hardly warrants

Above *Black Drop effect.*

Right *9.V.70 Transit of Mercury.*

their cost. If this method is employed, however, the difference between the sunspots and the disc of Mercury will be very apparent. It must be remembered that though the sunspots mark cooler areas of the Sun they are still very hot and luminous. It is the pure contrast that makes them appear dark. Mercury is at this time in no way luminous, so the difference in intensity between the planet and the spots is considerable. The transit of 1970 gave an ideal opportunity to observe this effect. On this occasion Mercury passed directly over a fairly large sunspot and the difference could easily be seen, even with the projection method.

At times, the Sun will go through a quiescent period and no spots will be visible at all. The image of Mercury will then stand out in solitary isolation as the only blemish on an otherwise clear solar disc. At such times no experience whatsoever will be needed to pick it out. This was the case in 1973. The illustration presented here is of the 1970 transit. There are a couple of sunspots on the area of Sun shown but the image of Mercury is unmistakable.

The transit can be divided into the four stages listed below. Timing them is very useful but can be difficult.

a) First contact of Mercury's preceding limb with the solar limb.
b) Moment of complete entry of the planet on to the solar disc or second contact.
c) Contact of the preceding limb of the planet with the solar limb near end of transit, or third contact.
d) Final disappearance of the planet's following limb from the solar disc, or fourth contact.

a) The problem here is the uncertainty of the point of entry. By the

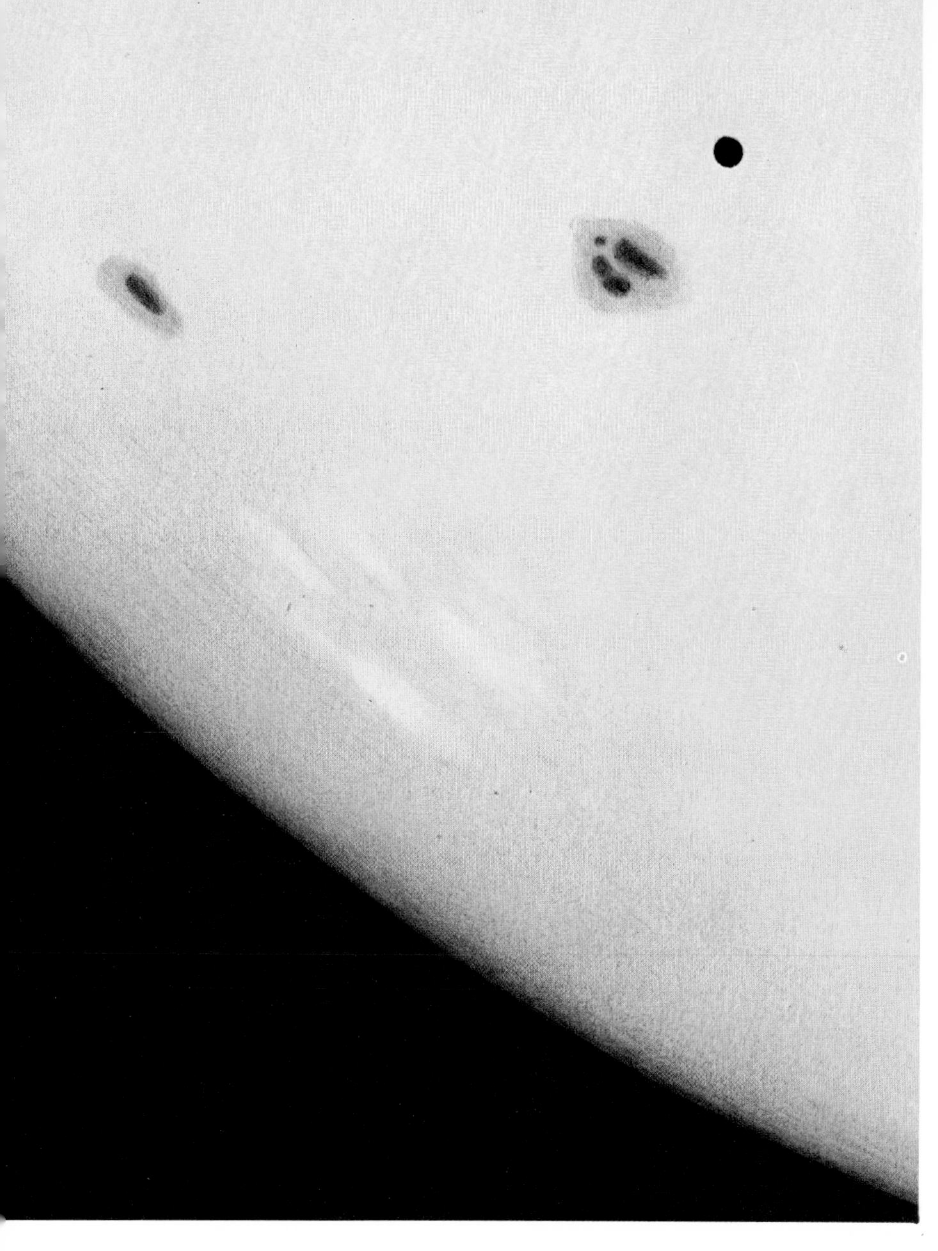

time the observer notices any sign of a notch the transit may already be well advanced.

b) and c) Difficulties arise because of the effect known as the 'black drop'. As the planet passes fully on to the Sun's disc one expects to see a line of light behind the planet as immersion is completed, but this does not appear immediately. A dark filament or tail seems to join the planet to the limb some considerable time after the theoretical second contact. By the time this disappears the planet is already well on to the Sun's disc. The result is an inaccurate timing. Similarly, before the planet reaches the limb of the Sun for third contact, it will be joined by a dark filament prior to the true contact. Thus accurate timing is again virtually impossible. It is now believed that the effect is probably due to the atmosphere of Earth and the resulting loss of sharpness, affecting both the planet and the Sun's limb. Imperfect optics will also add to the effect of blurring the images, allowing them apparently to merge before third contact or remain joined after second contact.

d) This is perhaps the only timing that can be made with real accuracy.

There have been many strange effects recorded while the planet is actually in transit, most of which are brought about by contrast or optical effects. Bright or dark haloes have been seen surrounding the planet, giving the appearance of an extensive atmosphere. Bright points of light on the black disc of the planet have also been seen, sometimes appearing central and at other times displaced considerably from the centre. It would now seem that each of these phenomena can be made either more or less obvious by certain optical combinations and by varying the magnification used with a given telescope. No rule can be laid down that will fully account for the effects, but the general opinion is that they are instrumental in origin. This is again something that can only be resolved by observation and experimentation.

If it were possible for us to pay a visit to Mercury the scenery would strongly resemble that of the Moon and, from the surface, the horizon would only be about four kilometres away. The Sun would appear to vary in its angular size, owing to the planet's eccentric orbit. At aphelion it would appear just over twice the diameter we see it from Earth, but at perihelion at least half as large again. The motion of the Sun in the sky of Mercury would seem very slow indeed when compared to its motion as seen here on Earth. To an observer at Mercury's equator, there would be 88 Earth days from sunrise to sunset and equally a night would also last 88 Earth days. So a full day and night on the planet would last in total 176 Earth days, or two Mercurian years, even though its axial rotation period is only 58.6 Earth days. Even more strange, during one of its days the stars will have travelled the sky three times since, relative to them, Mercury will have completed three rotations.

This is just a peculiar result of the relation between its rotation period and its sidereal period. Having no appreciable atmosphere would mean that the sky would be permanently dark, even with the Sun above the horizon. There would, of course, be no Moon in the sky but two other bodies would be very conspicuous. Venus would be extremely bright since, at its closest, it would present a full disc of 70″ in diamater; the Earth would be about as bright as Venus appears to us. Mars would look fairly bright while Jupiter and Saturn would look as they do from the Earth.

The extremes of hot and cold on Mercury are quite astounding. At the subsolar point the temperature would be as high as 400°C. Any atmosphere present would not be capable of retaining heat and the temperature would drop very rapidly after sunset to around −170°C at its lowest. These sudden changes from very hot to very cold would doubtless have an effect on the terrain, with rocks being cracked or broken by continual expansion and contraction. All in all, both observationally and physically, Mercury seems a very unsociable little world.

Venus

Venus, like Mercury, is an inferior planet, but here the resemblance ends. Take visibility, for example. Unlike Mercury, Venus is by no means difficult to identify. In fact it is one of the most obvious of the celestial objects. The reasons for its extreme brilliance are two-fold. Venus can approach the Earth closer than any other planet. Its distance from the Sun is 108.2 million kilometres and it can at times come to within 40 million kilometres of the Earth. Admittedly, the planet is on these occasions at inferior conjunction and is not easily visible, but for the six weeks preceding and following inferior conjunction it is still close to us. The second, and more important, fact is that Venus is covered by a dense atmosphere composed almost entirely of carbon dioxide, with some nitrogen. Dense opaque yellowish sulphurous clouds feature in the planet's upper atmosphere, and these together with the atmosphere act as a very good reflector of sunlight. In fact, 75% of the sunlight Venus receives is reflected back into space and this high albedo results in the very brilliant object we see in our dawn or evening sky.

The actual distance between Earth and Venus varies greatly as the two planets orbit the Sun. At superior conjunction, it can be as much as 257.4 million kilometres, causing the apparent diameter to alter considerably. At superior conjunction, it is no more than 10″ and yet at inferior conjunction it may be as much as 63″. Even so, the brightness only changes by just over one magnitude, i.e. −3.2 to −4.5, because the percentage of sunlit area visible to us decreases as the size of the disc increases. The planet is at its most brilliant 35 days or so before, or after, inferior conjunction, when the phase is about 27% and the diameter approximately 38″. Venus is often referred to as our 'sister world' but its only qualification is its actual diameter which is 12,100 kilometres, only 656 kilometres less than that of the Earth.

Since Venus's greatest elongation distance from the Sun can be as much as 47° there are times when it is very favourably situated for observation

Sulphurous clouds cut down visibility on Venus's surface to just a few miles.

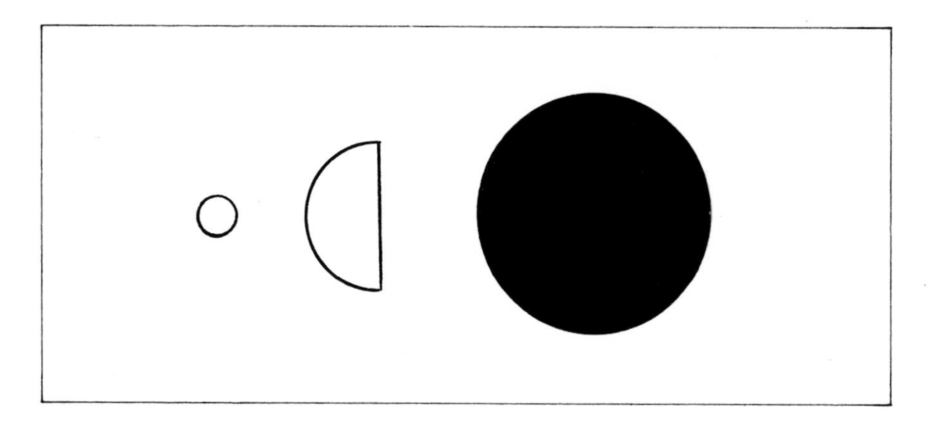

Venus's image size as seen at certain phases.

and may set well over four hours after the Sun. Not only can the planet be seen against a dark sky but it can often be spotted above the horizon near midnight. At this time, though, it would be too low for useful observation. For an evening elongation, an observing programme should begin as soon as possible after superior conjunction, say when the disc has attained a diameter of 15″, and should continue until the illuminated portion of the disc is too small to be of use. For morning elongations, begin observing as soon as the crescent is thick enough and continue at least until the diameter has shrunk to 15″ that is, approximately midway between elongation and superior conjunction.

Extreme brilliance

Owing to the extreme brilliance of the planet it is not practical to observe it when the sky is dark. The glare is then so considerable as to render the faint shadings to be seen on the planet quite invisible. Venus, because of its brightness, is easily visible to the unaided eye while the sky is still light and with the Sun still a little way above the horizon. These are, in fact, the best times for observation. The brightness of the sky acts as a filter in cutting down the glare so that the planet appears sharp and the contrast is good. Experience has taught us that the best time to observe is, as in the case of Mercury, during the two hours prior to sunset, or following sunrise, providing the altitude of the planet is sufficient. Seeing is usually very good during this period, with the sky brightness somewhat less than full daylight, allowing rather better contrast. The procedure for finding Venus during the day is the same as for Mercury except that it is much easier to locate, because of its brilliance, and can, with a good clear sky, often be picked out at mid-day without optical aid.

It has been claimed that the crescent of Venus can be made out with the unaided eye by keen-sighted observers, but the author has not been able to discern it despite many attempts. Still it is a fact that when the planet is between greatest elongation and inferior conjunction the smallest amount of magnification will reveal its shape. A pair of 7 × 50 binoculars will make the planet quite large at such times. Do not, however, use binoculars on Venus with the Sun above the horizon—it can be risky.

You will not need a large telescope for useful and worthwhile observation of Venus. Even a 75-mm refractor will suffice, although larger instruments are preferable and a 150-mm reflector will give much better results. The ideal size would be a reflector in the range of 200 mm to 300 mm. Unfortunately, Venus does not bear up well to high magnifications. The contrast of the markings is always very low and over-magnification will only make matters worse. A good guide is to use about ×1 for every millimetre of aperture, for example, ×250 for an aperture of 250 mm. There will, of course, be the odd occasion when a higher magnification can be employed just as there will be times when the recommended magnification is too high. The observer himself will be the best judge of this, and the foregoing is only intended as a rough guide. But remember, it is the contrast that is important, not a large image. The only time when a greater power becomes essential is when the planet is on the far side of the Sun and is exhibiting a small disc. You will find then that the amount of useful work that can be done is considerably reduced because of the need for finer definition.

As with Mercury one of the most useful contributions an amateur can make is to determine the phase percentage and the date of dichotomy of Venus. These tend to be much more difficult with Venus and the scatter of observers' estimated dates is often far greater than for Mercury. There are a number of reasons for this. First, the changing phase of Venus is a much slower process than for Mercury—the swift motion of Mercury means that the phase alters noticeably each night, even at greatest elongation, making the date of dichotomy easier to assess. It may also be a result of Venus's dense atmosphere, since there is a considerable fall-off of light towards the terminator, which adds to the problem.

In this respect the circumstances affecting Mercury—the conditions of the sky, whether hazy or clear—also apply here. If the sky is too dark and the image of Venus very bright then the phase will appear too great, due to 'irradiation'. Different magnifications employed will probably result in different values. Therefore, once a good average magnification has been found for a given telescope, it should be preserved, so that consistent results may be achieved. The coating on the mirror in the telescope, whether fresh, or old and dull, could affect the result. One should not be put off by this, however, because, as with Mercury, the more observations there are when averaged out, the better the results will be.

It is, of course, worth trying for accurate results—opportunities for observation may be very limited at a particular time. Bad weather in other parts of the country could, for example, mean that other observers have not seen the planet recently. Your observations will then be of vital importance in filling a gap and will if inaccurate, distort final results.

29.1.77 15.40 UT 419mm Refl × 250 Markings are fuzzy and ill-defined. The planet is shrouded in yellowish cloud.

In order to get less scatter in their results, the observing body of the British Astronomical Association concerned with Venus have decided to adopt the use of a Wratten-15 yellow filter. This was first done in the latter part of the 1960s and, with Venus being viewed through this filter, the spread of results from various observers has been cut down considerably. In this way, the observed phase seems to match the theoretical one much more closely. The filter certainly improves the steadiness and contrast of the view, since atmospheric turbulence has less effect on the small pass-band of light when seen through the filter.

Results that were obtained show that dichotomy occurs on average six days early at eastern elongations and four days late at western elongations.

Below *17.VI.75 20.30 UT 254mm Refl ×220* Right *21.I.77 16.00 UT 419mm Refl ×350 The Schröter effect.* Below *On the predicted date of Dichotomy the observed phase is well under half.* Right *shows the phase as half although the predicted date for Dichotomy was 26 January.*

In the past, from the time when the effect was first noticed by J. Schröter in 1793, dichotomy has been out by as much as eight days. As far as the 'Schröter effect' is concerned, measurements made with an instrument known as a micrometer will be the most accurate. However, estimation by eye alone, and measurement of drawings can give good results.

Some drawings presented here show the appearance of Venus around the time of dichotomy. The drawing of June 17, 1975, was made on the predicted date of dichotomy for the particular elongation and, as can be seen, the observed phase is well under half. The drawing of January 21, 1977, shows the phase exactly half, five days before its predicted date. Both these drawings were made during evening elongations and the trend of early dichotomy is well illustrated.

Dense atmosphere

Very little in the way of definite markings can be made out on the planet in visual light. The dense atmosphere seems to be more or less clear of dark patches or streaks. With some patience, however, faint dusky shadings, which tend to be very nebulous in appearance can be seen and, on rare occasions, faint brightish patches are recorded. It may be some time before an observer manages to detect any markings, particularly if the instrument used is a small one. New observers may follow the planet for several months and not see a single feature. Once markings have been detected, however, and one becomes familiar with their appearance, they are much easier to spot and it is unlikely that Venus will ever present a completely blank disc.

These markings fall into a rough pattern and often take the form of streaks, which are usually horizontal, sometimes diagonal and on rare occasions vertical, across the disc. They always seem more prominent near the terminator, which suggests some form of shadow phenomena, but this may be the result of reduced glare in the area, allowing the features to be more easily distinguished. Sometimes only a terminator shading is seen, which may sometimes be very extensive, spreading almost over the whole of the disc with the limb region bright and, apparently, the only part clear of shading. It is very rare in fact that

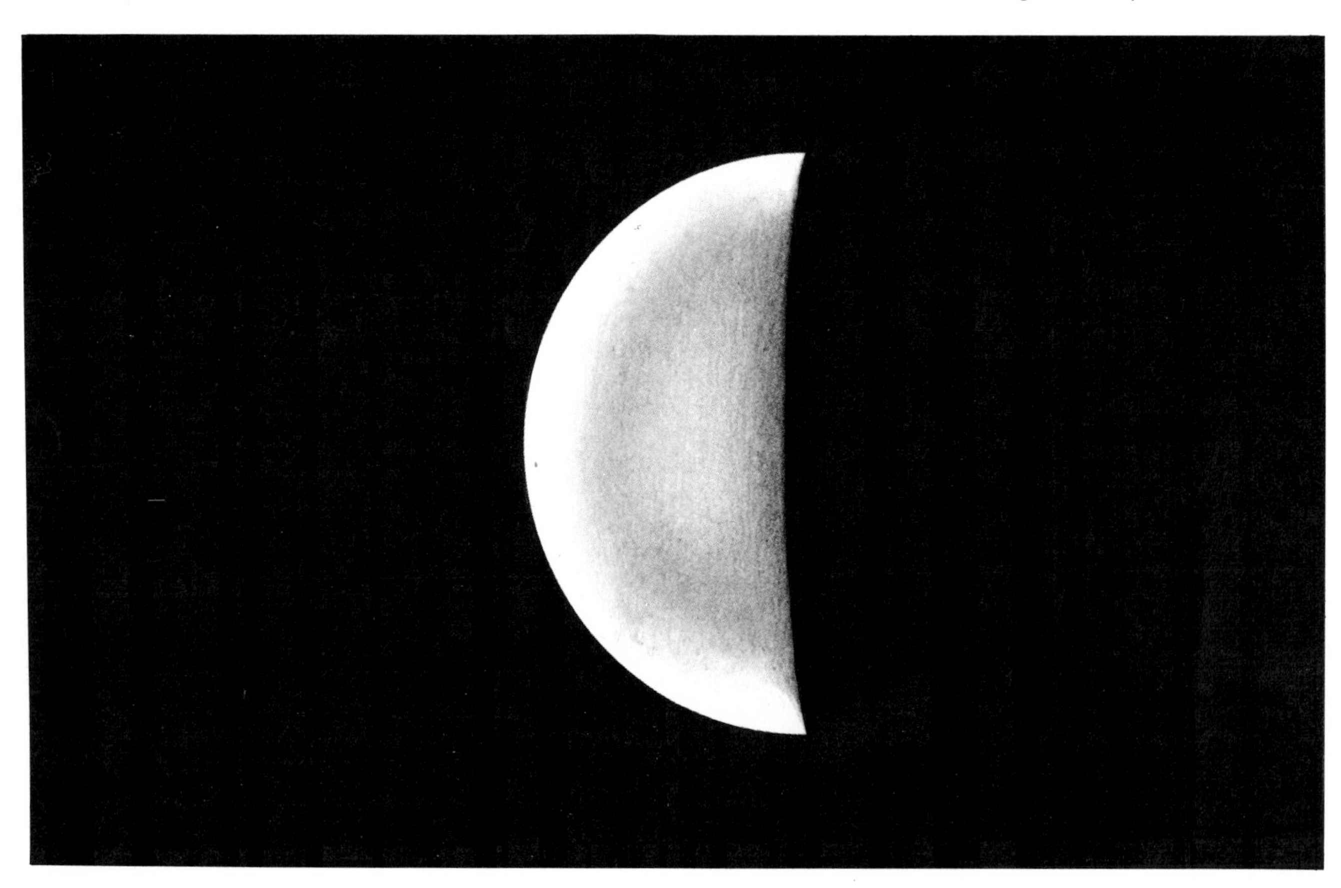

the limb region does not appear bright, whatever type of marking is displayed.

Sometimes a very dark patch, by Venusian standards, is recorded by a number of observers. Such an occasion arose in 1962, when a particularly dark patch was seen near the terminator. A further dark feature was recorded during May 1978 but on the whole they are very rare. Though not seen as frequently as the dark shadings, bright patches will, from time to time, feature on the planet perhaps several times through an elongation. They are never particularly striking but observations often agree on this type of feature. When these bright patches occur near the terminator they can be fairly prominent, often giving the appearance of bright bays along the terminator and, if bright enough, they can give rise to a slight irradiation effect making the terminator seem uneven. Sometimes a single very bright patch will be seen but, as with very dark features, this is rare.

By far the most common and well-observed feature on the planet is the appearance of bright areas at the cusps. These 'cusp caps', as they are called, are the only regularly recurring feature. At least one or the other will be noticed and commonly both will appear bright at the same time, giving the impression of polar caps. Until recently it was not known if these cusp caps were situated at the actual poles of the planet. In the light of information gained by radio observations and space probes, we have learned that the axial tilt is 178°. The planet is upside down in effect, its north pole being situated where its south pole should be. However, this tilt is still only 2° from vertical with respect to the plane of its orbit, so the cusps do, in fact, mark the positions of the true poles. The brightness in this region is an effect of the cloud formations in the atmosphere and is not caused by actual ice caps.

Information from the Pioneer Venus space probe that reached Venus in December 1978 has told us a great deal about the planet's clouds. It seems that there is a ring of high cold cloud surrounding each pole. Furthermore, a haze, surrounding the planet, thickens over the poles and becomes opaque to long-wavelength radiation. It is thought that this could result from ice crystals in this polar haze and a comparison may be drawn between these and the cirrus clouds found over the polar regions of the Earth. This suggests that we are seeing real features in the cusp caps brought about by cloud layers with higher reflectivity than the general cloud layer.

The actual rotation period of Venus has been found to be 243 days, which is longer than its year. It is also retrograde, this being a direct result of the fact that it is upside down. The dense atmosphere had made determination of the rotation virtually impossible and it was only the visiting space probes and radio observations that gave us this information. The rotation of Venus's atmosphere is a different story. This was determined visually, that is, photographically, from the Earth.

Photographs in ordinary white light show very little in the way of markings but photographs taken in ultra violet light show very conspicuous markings. It was with these that in the late 1960s the four-day retrograde rotation of the atmosphere was discovered. It is sometimes possible to detect signs of this period in one's own drawings and the period is quite

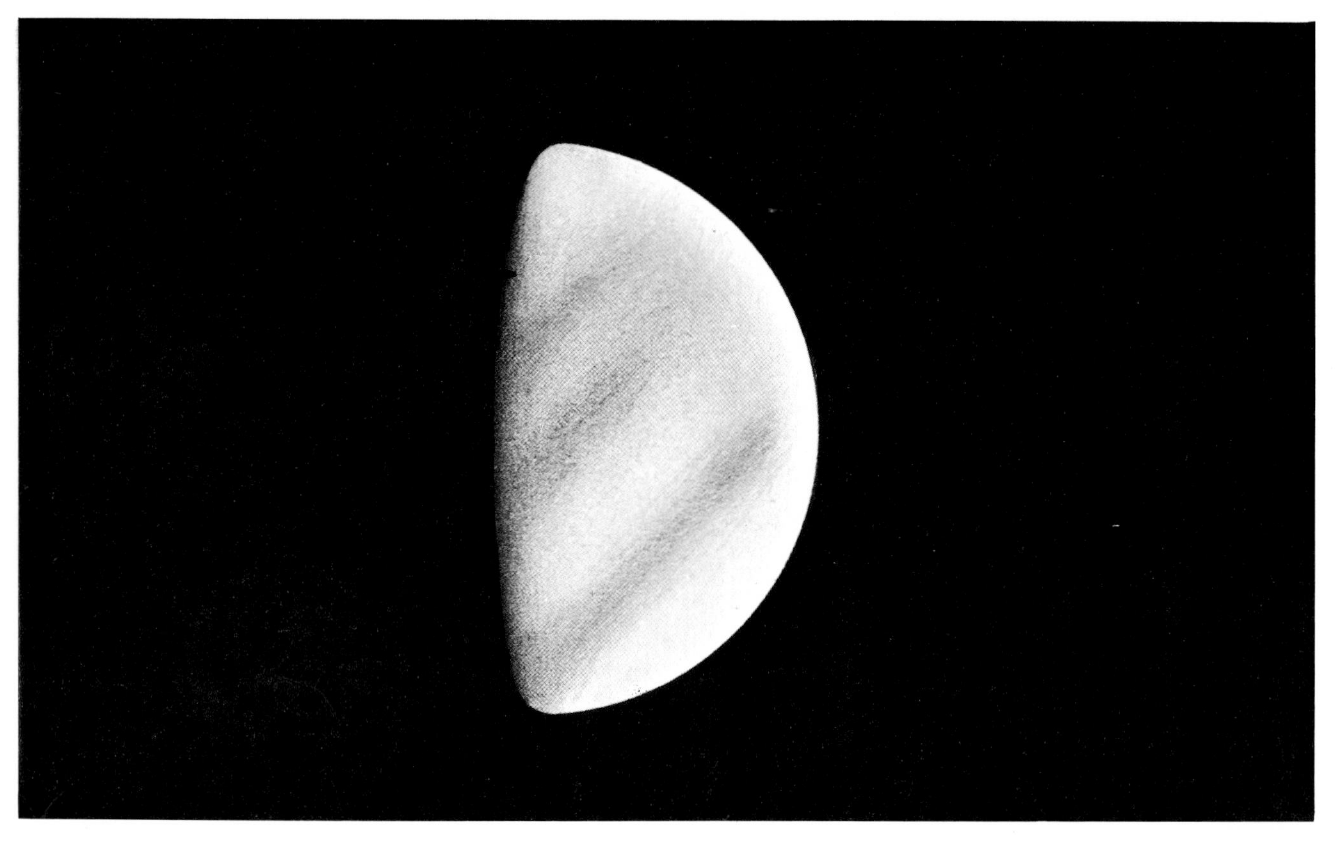

Above *30.XI.75 10.35 UT 254mm Refl ×280* Right *8.XII.75 10.40 UT 254mm Refl ×300 Evidence of the four-day rotation period of Venus's atmosphere.*

marked when a number of observations from different observers are considered collectively. The difficulty of establishing this, however, is illustrated by the fact that, despite the observational coverage the planet has received since the invention of the telescope in 1609 and its subsequent use for astronomical observation in 1610, the period was not recognized until the middle of the late 1960s. It is a fine example of how useful visual observation can be. Two of the drawings presented here show a recurrence of markings on drawings made eight days apart.

Another feature observed quite regularly on the planet is what appear to be dark borders, or collars, surrounding the bright cusp caps. It may be significant that these features always appear together, though it is not as yet fully established whether they are the result of pure contrast. Mariner pictures do, however, show darkish belts adjacent to the polar regions, confirming observation.

It is useful to try to record the relative intensity of the various features seen. Drawings should be made, whenever possible, to supplement this work, and the region that a particular intensity refers to indicated. A standard scale of intensity should be used throughout and the one adopted by the British Astronomical Association is as good as any. This is as follows:

0 Very bright areas (rare)
1 General hue of planet
2 Markings on the limit of visibility
3 Dusky shadings still rather difficult
4 Shadings easy to see (rare)
5 Very dark areas (practically non-existent)

Intermediate values may also be used if necessary, i.e. 2.5. With a little practice this method will become easy, forming a valuable part of one's records. Evaluation of estimates supplied by a large number of observers should make apparent any unusual feature.

Apart from the regular features there are certain strange phenomena observed from time to time. Irregularities in the terminator, taking the form of dents, are often reported, though in general these are seen with smaller instruments and usually when seeing conditions are poor. It is likely that in most cases this is a spurious effect. From time to time, however, they are reported by reputable observers using large telescopes. It may be that a particularly dark marking on the terminator is responsible, or large cloud areas in the vicinity of the terminator could give rise to shadow phenomena which would probably have the same effect.

Strange phenomenon

More unusual still are the occasional reports of small bright points of light seen beyond the terminator, well into the night hemisphere. Again, J. Schröter featured strongly in the early reports of these and as with Mercury the high mountain explanation was put forward. This idea can, of course, be disregarded but even in recent times such phenomena have been reported by trustworthy observers and if anything of the sort is seen it should be reported at once in order to gain confirmation. Another unusual aspect that the planet presents is an apparent difference in the shape of the cusps, as was the case with Mercury. With such a considerable atmo-

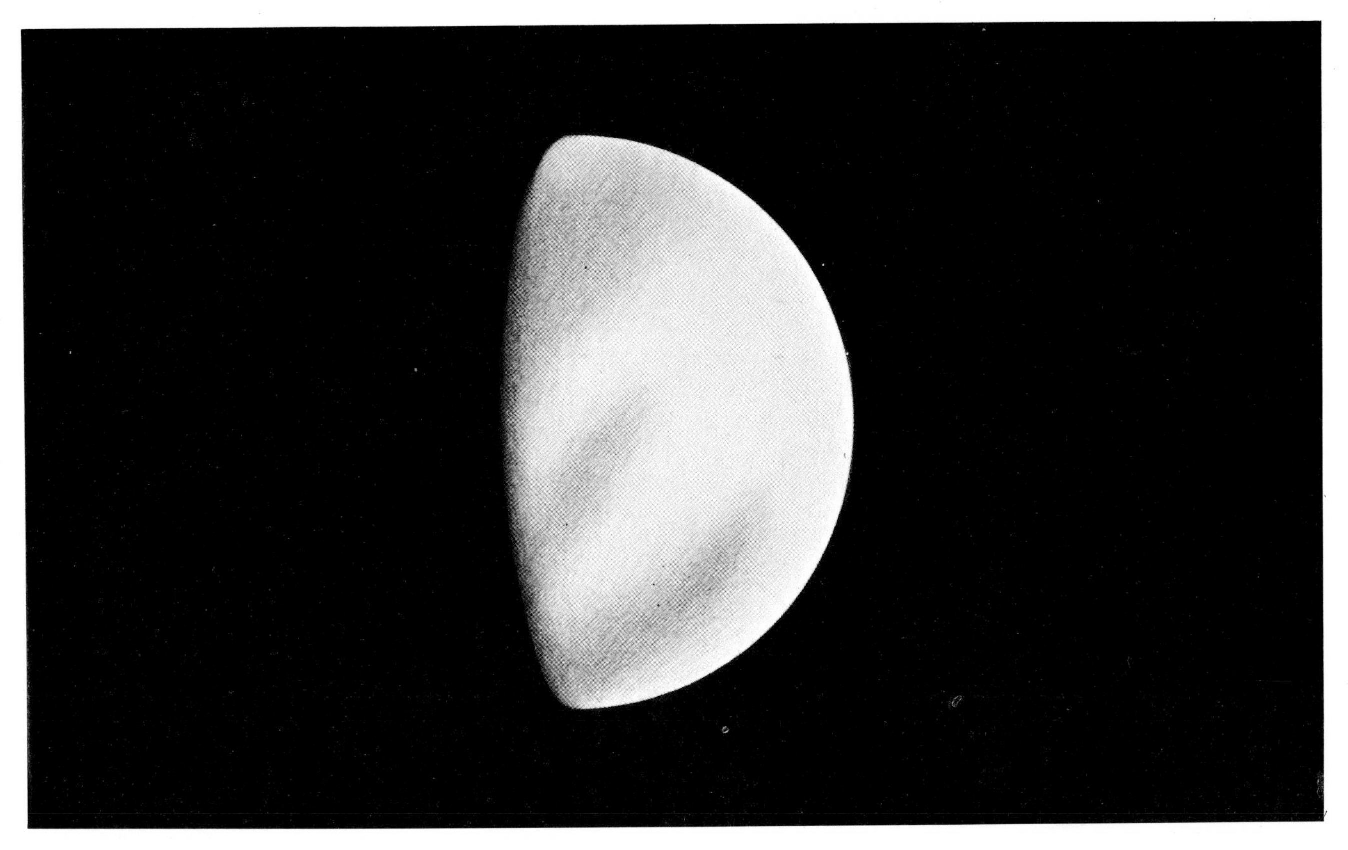

sphere present this type of effect is not unexpected.

One of the most difficult observations to make is to detect a faint glow of light on the night side of the planet, which is reported from time to time. This resembles the effect of Earthshine seen on the Moon. In the case of Venus, there is, of course, no second body to bring about this effect and the planet's atmosphere must be responsible in some way. The light received from Earth is certainly not sufficient to be reflected back, even though it would be an exceptionally bright object as seen from Venus. The phenomenon is known as 'ashen light' and it is generally accepted as a real feature. But even when it is seen, it is always on the very limit of visibility. Theories for its cause have been numerous, but the favourite one is that it results from the streams of electrified particles emitted by the Sun striking the upper atmosphere of Venus and exciting it into luminosity. This is a familiar effect here on Earth and is known as the 'Aurora'. Venus is, of course, closer to the Sun than we are, so it is reasonable to assume that any such effect would be of a greater intensity than that experienced on Earth. Whether this effect would be observable from the Earth, however, is questionable.

Maintaining a continual watch for ashen light is essential if we are to resolve the matter. A large number of observations spread over many years will, for instance, tell us if the effect is more frequent around the times of sunspot maximum. This is the time when our own Aurorae occur most frequently. You should watch out for the effect particularly when reports of intense solar flares are received. The best time to attempt observations is when the crescent of Venus is rather thin, i.e. below 30% illuminated. The sky also needs to be quite dark, unlike other observations of the planet, and the sky should, above all, be crystal clear. The slightest haze will cut down transparency sufficiently to render so faint a feature completely invisible.

A very useful and effective aid to this type of observation is an occulting bar, or something that blots out only the bright part of the planet and reduces the glare and stray light. This need only be a thin piece of wire with a U-shaped bend in the middle, which is placed in the focus of the eye-piece of the telescope. Fuse wire will usually be thin enough and the correct U-shape will be found by simple trial and error. With the bright portion of the planet eclipsed, so to speak, the chances of spotting ashen light will be much improved. Look for the effect about three weeks before, or after inferior conjunction, but do not expect it to be easy, because it won't be.

Taking care

Observations of Venus when it is at or near inferior conjunction can be very rewarding. Great care must be taken when making such observations, because the planet is then close to the Sun and observers without some experience should not attempt them. Venus has an orbit that is inclined 3° 23′ to the plane of the ecliptic and, though only a small inclination, it is the third greatest after Pluto and Mercury. Only on very rare occasions when inferior conjunction occurs at the nodes (which will be discussed later) does the planet pass directly in front of the Solar disc as seen from Earth. Usually as a result of its inclination it will pass either north or south of the Sun, sometimes by as much as 9°. The planet may then be located with comparative ease and the sight can be very dramatic. Its diameter appears at its greatest, 63″ and all that is seen is a delicate slender crescent that resembles a fine piece of silver wire.

The extent of the crescent may be greater than a semi-circle, with the cusps being traced most of the way round, giving the appearance of a complete ring. In 1975, for example, with the planet 8° south of the Sun at inferior conjunction, the appearance was not that of a complete ring, but the cusps were considerably extended and formed at least 70% of a circle. The reason why Venus can sometimes be seen as a ring is yet another effect attributable to its atmosphere. This dense cover refracts the sunlight some way round into the dark hemisphere. To see this easily, conjunctions must be closer than 8°. Again, crystal clear skies are necessary for successful observations. The occasions when Venus displays cusp extensions are not always limited to times of inferior conjunction. In fact, slight extensions are recorded when the phase is a thick crescent or even half. Bright auroral activity has been suggested as an answer to the phenomena but observations are far too infrequent and erratic to allow any sound conclusion.

Transits

Transits of Venus across the face of the Sun are extremely rare—the last occasion was in 1882. At present, transits are occurring in pairs separated by a period of eight years. A situation exists with Venus where a cycle of phases will be repeated, almost exactly, every eight years, the only difference being that this cycle starts two and a half days earlier each time. For example, the inferior conjunction of 1972 occurred on June 17. That of 1980 will occur on June 15. The transit of December 6, 1882, was preceded by another on December 9, 1874. We will unfortunately have to wait some time for the next pair since these will take place on June 8, 2004, and June 6, 2012.

It is understandable that the silhouetted disc of Venus appears very much larger than that of Mercury since the planet is itself larger and also closer to us. It can in fact be seen without optical aid through a piece of dark sun filter. Information given for Mercury regarding the difficulty of obtaining timings for the event will apply here also. They are, in fact, harder to obtain, the main problem being the black drop effect, which is much more evident in the case of Venus, perhaps because of the planet's own considerable atmosphere. This makes the timing of second and third contact a practical impossibility. Adding to the problem is the much slower apparent motion of the planet, creating larger errors in timing. Despite the difficulties, the sight of the planet in transit should be, to say the least, spectacular. It is a pity that we have to wait so long for the next opportunity to witness this.

The colour of Venus

Venus does not exhibit much in the way of colour but descriptions of what colour is present often differ. To the author it usually appears pale yellowish-white with faint grey shadings and occasional almost white patches. If both Venus and Mercury are situated so as to be visible in the same telescopic field, the difference of colour between the two is striking. In 1961 they were both visible to the author in the field of a small telescope with a low power. The brilliant silvery yellow of Venus contrasted strongly with the dull warm pink of Mercury and the sight was very beautiful. We now know that there are sulphurous clouds in Venus' atmosphere and these no doubt give it the yellowish colour seen during daylight.

It is, of course, the space probes that have taught us so much about the physical make up of Venus' atmosphere and the appearance of its surface. The first successful Venus probe was Mariner 2, which made a close approach to the planet in 1962. This confirmed the planet's 243-day retrograde rotation and indicated that surface temperatures are very high. Mariner 5 followed in 1967, improving knowledge a little, but the highly successful Mariner 10 spacecraft was the first to send highly detailed photographs of the planet's upper atmospheric cloud patterns. The prime target for this craft was really Mercury but a close approach to Venus in February 1974 was used to swing it on to Mercury. The situation was used to advantage and a number of photograpns of Venus were taken as the craft passed. These showed considerable turbulence and convection currents in the equatorial regions, with high bright cirrus-type cloud formations in the polar regions. They also confirmed the four-day rotation period of the atmosphere. Stratification of the atmosphere was discovered together with the fact that the planet was covered by a considerable haze layer above its visible clouds. The photographs were taken in ultra violet wavelengths, which are also the only ones to show any reasonably clear markings when photographs are taken from Earth.

Several attempts to land on the planet were made by the Russians and though many of their probes did reach the surface, information received from them was very limited because of the very short active life of these probes, once on the surface. It appeared that the harsh conditions on the planet were too much for them. Real success came in October 1975, when the Veneras 9 and 10 actually sent the first pictures back from the surface. These showed a surface littered with rocks and surprisingly, despite the dense cloud layers, light at the surface was good, something akin to a dull cloudy day here on Earth. The areas shown looked distinctly volcanic.

Superb results have been obtained by the American spacecrafts Pioneer–Venus 1 and 2, which arrived at the planet in early December 1978. The mission involved both an orbital vehicle and also probes to penetrate the atmosphere. These probes have told us a great deal about the atmosphere and its temperature at various levels.

The first sign of a reasonable atmosphere was detected at an altitude of 250 kilometres above the surface. A dense sulphurous haze was encountered at an altitude of 70 kilometres, where the temperature was around 2°C and the atmospheric pressure about half that of Earth's atmosphere at sea level. This was the first of the atmospheric layers that reflect most of the sunlight received, giving Venus its brilliance and yellowish coloration.

Slightly lower, at 60 kilometres' altitude, a denser haze layer was encountered, containing sulphuric acid and sulphur particles. The temperature here was around 40°C. Following a slight clearing, a third layer was discovered, which was far denser than the previous layers and laden with clouds. These clouds contained much more sulphur and the temperatures here were around 80°C with atmospheric pressure equal to that of the Earth at sea level. The altitude was around 50 kilometres. At 30 kilometres' altitude, the atmosphere seemed to be fairly clear with the possibility of good visibility. Finally, at the surface, temperatures were around 450°C and atmospheric pressure could be as high as 90 to 100 times the Earth's at sea level. Temperatures at the surface are so high because of what is known as the 'greenhouse effect'. Heat radiation from the Sun penetrates through the clouds but is trapped beneath them. The heat then builds up to this phenomenal level. Only one of the probes succeeded in transmitting data from the surface and even then for a period of only 67 minutes. Its instruments evidently could not stand up to the harsh conditions for longer.

Later in December of the same year two more Russian probes succeeded in landing on the planet, Venera 12 on December 21 and Venera 11 on December 25. They transmitted data from the surface for 1 hour and 50 minutes and 1 hour, 35 minutes respectively. No photographs were returned by these probes but one interesting thing was the detection of thunderstorms. A discharge in the vicinity of one of the probes caused vibrations which lasted for a quarter of an hour.

Imaginary view

Despite all of this information it is difficult to imagine what a view at the surface would actually be like—very different from anything we are used to. Visibility might be as low as 3 kilometres and the light probably dull and reddish. This might possibly result from dust suspended in the atmosphere. The sky would obviously be heavily clouded with probably no sign of the Sun whatsoever (not unlike England sometimes!). The ground, though fairly flat in the region of the probes, would be covered with rocks and there would be a fair amount of dust about. The sulphuric acid droplets in the clouds would be very corrosive.

We have a lot of information about the atmosphere and a couple of very localized views from the surface but, because of the dense cloud cover, we have no wide angle views of the planet's surface from an orbital vantage point. Information of what the general surface may look like was obtained by radar—in fact, by bouncing radio signals off the planet. This has revealed a very rough terrain.

The most interesting feature recorded is a large canyon which seems to be in the order of 1,500 kilometres long by 150 kilometres wide and perhaps as much as 2 kilometres deep at its deepest. Other details have been recorded in this way, such as a vast number of craters that range in size from around 30 kilometres to a few thousand kilometres. Also, a large volcanic formation has been found with a diameter of 400 kilometres at the base and at least one kilometre high. At its summit there appears to be a crater of at least 80 kilometres in diameter.

28.VIII.75 10.20 UT 254 mm Refl ×200 At inferior conjunction the crescent can still be seen.

Theories about Venus

Altogether, Venus is a very uncomfortable world, which in no way resembles our previous ideas of the planet. Theories had ranged from a prehistoric landscape, monsters and all, to a planet covered with oceans of soda water; or a scene of violent volcanic activity to one of a desert-type landscape, the latter, if anything being the closest to reality.

Although Venus does not present a telescopic object rich in detail and colour, there is much to repay the patience of any observer. Even to the casual observer the sight of the little silvery moon-like disc contrasted against the deep blue sky is ample reward for the effort of a search by daylight. You would never believe just by looking what an unfriendly environment exists on the planet!

Mars

Of all the planets, Mars can be one of the most rewarding to study if you intend to carry out serious work. Markings on the planet may at first be difficult to make out, but with experience the eye will come to detect the detail and the planet's appearance can be fascinating.

Mars is the first planet with an orbit outside that of the Earth, in order of distance from the Sun—in other words, the first superior planet. It has a fairly eccentric orbit with a mean distance from the Sun of 228 million kilometres. Because of this eccentricity, however, its distance can vary by as much as 53 million kilometres.

The planet is itself quite small. Its diameter of 6,790 kilometres is only a little over half the diameter of the Earth. The orbit of the Earth is fairly circular with the Sun quite centrally placed, so the eccentricity of Mars' orbit means that its opposition distance varies considerably, so that some oppositions are much more favourable than others. Oppositions occurring at perihelion can put Mars at a distance of only 56 million kilometres from Earth. At such times its angular diameter may reach 26″. Oppositions occurring at aphelion put Mars at a distance of 99 million kilometres from Earth, and the resulting angular diameter is no more than 14″. The greatest distance Mars can ever attain from Earth is at the time of superior conjunction. It can then be as much as 400 million kilometres away, with an angular diameter of only 3.5″. Obviously this is too small for useful observation and, in fact, the period of time when useful work can be carried out is rather short, being at best three to four months either side of opposition.

Observational problems

Since Mars' oppositions are spaced more than two years apart the situation is not very favourable. The most annoying fact, so far as northern hemisphere observers are concerned, is that the perihelic oppositions occur in August, when the planet is well south of the equator and so poorly placed from Britain. This means that when the planet is close and the angular diameter at its largest, we have to view it through a dense layer of atmosphere. Aphelic oppositions occur in February, when the planet is well north of the equator and thus riding high in our skies. When the

Right *Vikings' landers have given us an idea of the view from Mars. (Author's impression)*

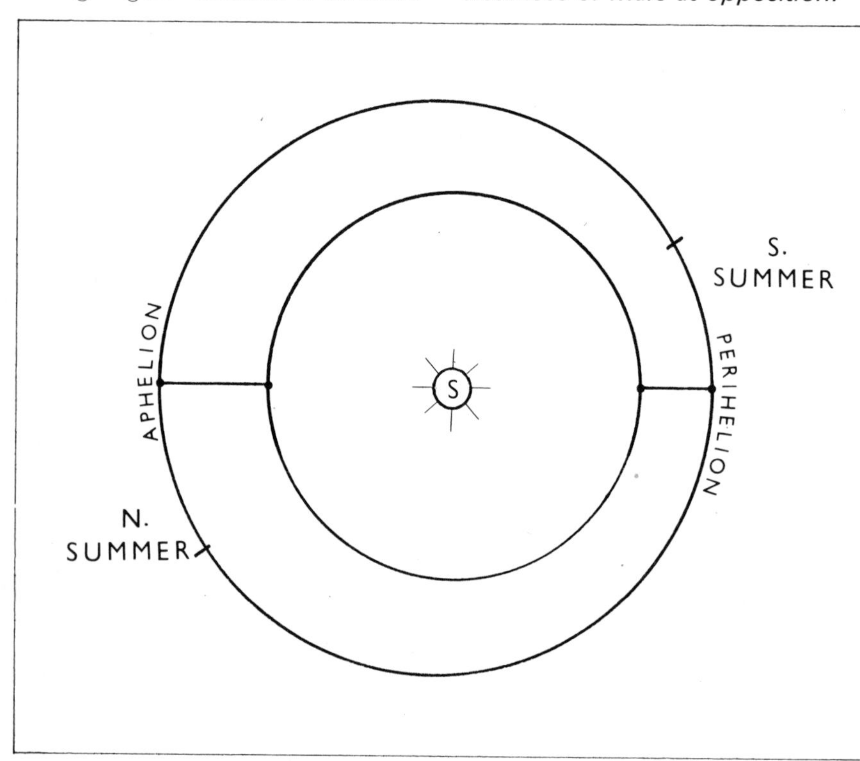

Diagram showing the varying distances of Mars at opposition.

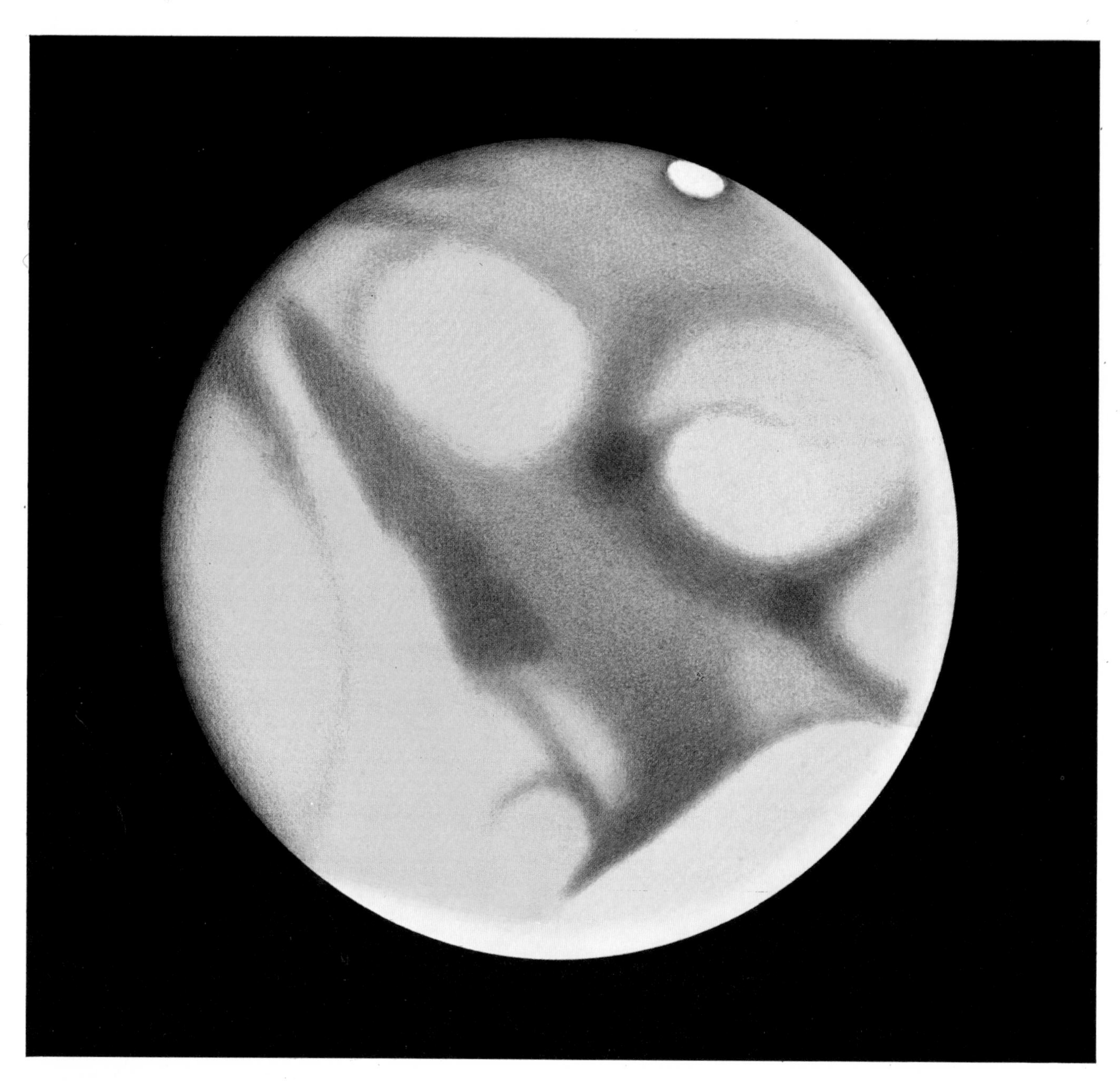

17.X.73 01.30 UT 254mm Refl ×380 Syrtis Major, a familiar feature of Mars—the first to be observed and recorded.

seeing is good, therefore, the diameter is small. There are rare occasions when northern observers may experience fairly favourably oppositions. Those occurring in October, for instance, find the planet still quite close to its perihelion point and also quite high in our sky, being situated on or near the celestial equator. Such an occasion arose in 1973 and the angular diameter then attained was 21″.

Being an outer planet Mars will never be seen to go through the full range of phases in the way that Mercury and Venus do. There are times, however, when a small portion of the unilluminated hemisphere becomes visible to us. Though this portion never exceeds 15% of the full disc, it does amount to a noticeable phase, giving Mars a gibbous appearance resembling the Moon when three days off full. Prior to opposition, this unilluminated portion will appear on the left of the disc as viewed through an inverting telescope. Following opposition, it will appear on the right. If accurate drawings of the planet are to be attempted then this phase must always be taken into account.

As far as the size of instrument necessary for useful observations is concerned, with Mars the bigger it is the better. Markings can be made out with small telescopes quite easily. In fact, most of the major features are visible with only a 75-mm refractor, if given the right conditions. It is also possible to do useful observing work with a reflector of 150 mm. In general, however, with this planet the refractor would seem to have a considerable advantage over the reflector, and a refractor of 150-mm aperture would allow very worthwhile work to be done. The problem here is the cost. Moderate-sized refractors are well beyond the price range of most amateurs. Colour work is never so accurate with a refractor, however, so at least this is one consolation to the reflector owner. Perhaps the ideal amateur instrument is a reflector in the 250-mm to 300-mm range. It must be admitted that the finest views are obtained with large instruments and if you can better this size then you should do so. Since Mars is a rather small object, fairly high magnifications will be needed. Something in the order of $\times 15$ per 10 mm of aperture is a good guide and, if ever conditions allow, it might be desirable to better even this. Unfortunately the planet often seems badly affected by seeing, more so than the other planets.

Mars has an axial tilt of 23° 27′, which is similar to the Earth's. As a result, the planet will display various tilts at different parts of its orbit. Summer solstice for its southern hemisphere occurs when the planet is close to the point of perihelion. This means that at the favourable oppositions occurring around the end of August the southern hemisphere will be tilted towards us. Our knowledge of Mars' southern hemisphere features is therefore far better than for the northern so far as our Earth-based observations are concerned. This latter portion of the planet is tilted towards us when Mars is near aphelion (around February). The tilt also means that Mars experiences seasons in pretty much the same way as the Earth. They are, of course, on average twice as long as our own, and more uneven in length, owing to the orbital eccentricity. A southern spring on Mars lasts for about 146 days, whereas its autumn lasts for 199 days. Also the southern summers, though shorter than the northern summers, tend to be hotter because the planet is closer to the Sun. The uneven length of seasons comes about simply because a planet with an eccentric orbit—no matter how slight—will move faster at perihelion than it will at aphelion. The higher the eccentricity, the greater the difference in orbital velocity between the two points.

The axial rotation period of Mars is known with considerable accuracy, and at 24 h 37 m 23 s, it is only 37 m 23 s longer than the Earth's. A well-established system of latitude and longitude is in common use and many astronomical handbooks, such as the one published by the British Astronomical Association, give a table of longitude for the Central Meridian of the disc for midnight of each day (Universal Time), covering a period of several months around each opposition. Also given is the amount by which the longitude changes over periods of time down to one minute.

The Central Meridian

With the aid of the table and a knowledge of the longitude change for given times it is possible to work out the actual longitude of the Central Meridian at any given time. An imaginary line running from the north pole to the south and passing through the centre of the disc forms the Central Meridian. By noting the time that a particular feature is situated on this line, one may derive the longitude of that feature. The fact that a Martian day is 37 m 23 s longer than our own day means that a given feature can be seen in the same position on the disc 37 m 23 s later each consecutive night. It also means that if the planet is observed at the same time each night over a period of about six weeks then, in effect, a slow retrograde rotation of the planet will be seen. Large portions of the planet's surface can be seen during a single observing period as the planet's rotation gradually brings different features into view. By waiting for a period of two hours the planet will present a completely different region and it is often possible to make three separate drawings in one night showing entirely different aspects. The main advantage of this is that maximum use may be made of a night when seeing conditions are particularly good. Furthermore, maximum coverage of each area on the planet can be obtained in this way and any unusual phenomena are likely to be noticed much sooner.

Unlike the atmosphere of Venus, that of Mars is usually quite transparent and we are able to see the actual surface in good detail. The features of the surface are now really well known and superbly accurate albedo maps have been produced.

The primary objective of observational work is to detect possible changes in the regular features. It is therefore essential that a good knowledge of the features is obtained. If it is necessary to refer to a map before an observation, in order to brush up on familiarity, this is permissible provided one can be disciplined enough to record only what is there, with no imaginary extras.

The actual features of Mars take the form of large dark areas and extensive bright regions. In the past, it was thought that the dark areas were seas and until recently the Latin names given to certain features reflected this, for example, Mare Tyrrhenum. They realized long ago that these so-called seas contained no water and generally believed that the dark markings were areas of low ground. There seemed, therefore, no reason why the names should not remain the same.

Martian features

As a result of close encounter photographs returned to Earth by various Mars probes, it was found that it is the bright regions that tend to be of low ground and the dark markings high ground. For this reason, it was considered necessary in some cases to change names. Mare Tyrrhenum is just one example. This has now become Tyrrhena Planum. Planum refers to a high plain. Some of the dark markings are very prominent, in particular a feature known as Syrtis Major, a large triangular-shaped area situated at Martian longitude 290°. It is possibly the most familiar region on the whole planet and it was even the feature shown on a drawing by Christiaan Huygens in 1659, which was the first to show a recognizable feature. To the south of what is now Syrtis Major Planitia, is a large circular bright region known as Hellas. This is a huge basin which, when centrally placed on the planet, forms with Syrtis Major its most familiar aspect. Both of these features are variable in size, shape and intensity and are typical in this respect of most features on the planet. At first glance, Syrtis Major will appear uniformly dark but, when studied in excellent conditions of seeing with high magnification, a mass of detail will reveal that the whole thing seems to be made up of numerous spots and streaks.

Unfortunately, capturing such detail on a drawing or photograph is seldom possible. Again, this appearance is typical of most of the Martian features. There are, of course, other well-known areas which will eventually become familiar landmarks to new observers, for example, the Sinus Sabaeus and Sinus Meridiani region. It is, in fact, the forked bay of Sinus Meridiani, that marks the prime meridian of Mars (0° long). Some of the drawings presented here show familiar aspects.

Beautiful spectacle

It is the colour of Mars—obvious even without optical aid—that is the most striking of all its characteristics. The sight viewed through a decent-sized telescope it is magnificent. The general colour of the disc is fiery orange marked with patches of dark greenish, or bluish, grey. These may at times appear so dark as to match the intensity of the background sky. Glistening white polar caps may be seen at either pole, akin to our own great polar ice caps. They complete a very beautiful picture when advanced in their development. Seen in large telescopes the spectacle is unbelievably beautiful. The basic bright reddish areas will be resolved into regions of pink, yellow, orange and red, some parts having a dull greyish cast while others are bright and clear. The dark markings, their colours quite possibly enhanced by contrast with the reds, will appear equally patchy and varied in colour, some being bluish grey, others having a greenish tinge. Some parts will appear greenish brown while odd patches will be brownish grey. The polar caps on first impression, look stark white, but if both visible, one may have a yellowish tint while the other appears bluish white. To see such colour, the conditions must, of course, be excellent. The overall picture is one of bright reddish orange areas to the north and extensive bluish grey markings to the south. What bright features there are in the south tend to be fairly isolated as do most northern dark features.

No description of Mars would be complete without some mention of the famous canals. It was in the latter part of the 19th Century that the celebrated astronomer Giovanni Schiaparelli observed what he took to be linear features criss-crossing the planet's bright regions. He announced his observations of these in 1877 and referred to them as *canali*, or channels, but indicated at the time that he believed the features to be natural. The term was taken to mean canals, however, and from this they went on to assume that they were artificial and built by intelligent Martian inhabitants. It was a nice idea and one which was followed up by a number of astronomers, among them Percival Lowell. Although Schiaparelli first drew attention to them, it was Lowell who gave them the publicity they most certainly did not deserve. Lowell imagined the Martians building enormous canals to irrigate their dry and dying planet with water from the polar ice caps. (No doubt this idea has formed the basis to a great many science fiction stories.)

5.X.73 22.30 UT 254mm Refl ×380 The dark Sinus Sabaeus.

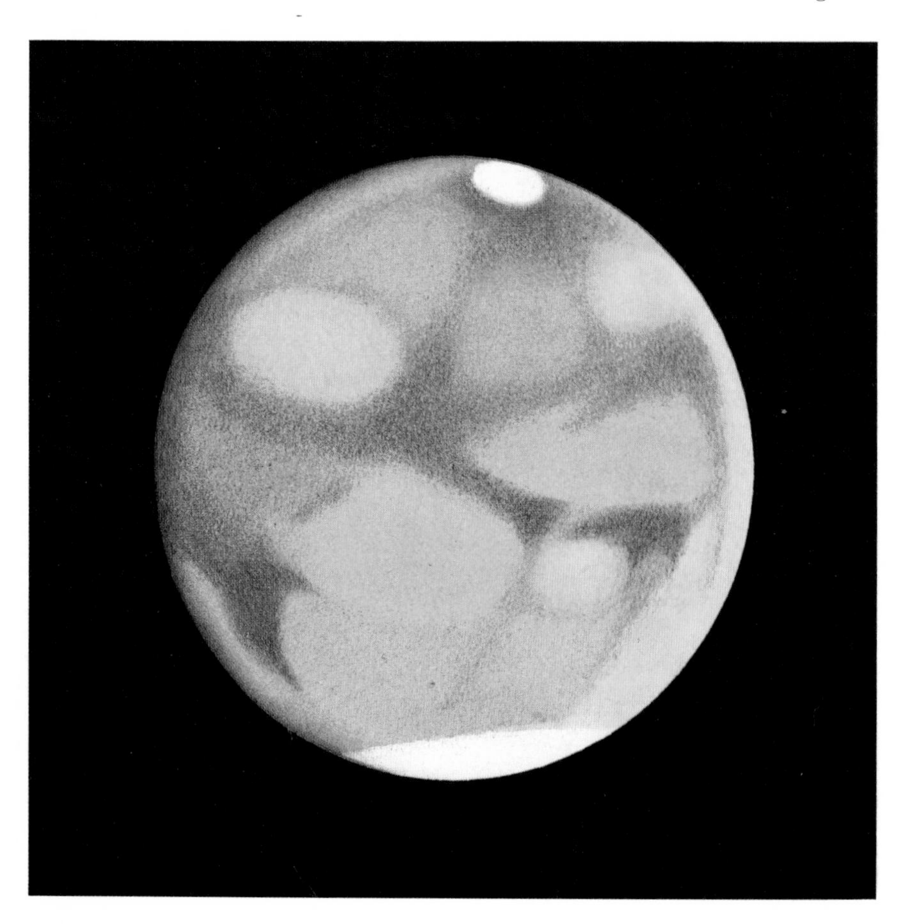

Theory disproved

To prove his point, Lowell undertook the mapping of these canals with the 600-mm refractor at Flagstaff, with the result that his finished work resembled a spider's web. The idea that the canals were artificial was strengthened by his observations, which later indicated that some of the lines became double over certain periods and looked like train lines. It was also noticed that with the shrinking of the polar cap during the Martian spring and summer, a so-called 'wave of darkening' seemed to sweep slowly over the planet from the pole to the equator, the canals and dark areas surrounding them gradually becoming darker as the summer approached. It was as if the large dark areas were vast regions of vegetation, refreshed by moisture from the melting caps, and the canals were designed to aid in this process. Lowell was entirely convinced that these features were real, even though many other eminent astronomers failed to see them. He explained this away by suggesting that the conditions where his observatory was situated were far superior to the conditions at other observatory sites, which was to a large extent true at the time. Lowell himself had been involved in choosing the site and had taken great care to ensure that it was good.

Certainly there are crude linear-type features on the planet, but these are a long way from the narrow sharp lines seen by Lowell. Any line feature seen from Earth would have to be very wide in reality—something like 100 kilometres, which would make them very wide canals indeed. Moreover, in the finest conditions and with telescopes far superior to the one used by Lowell astronomers have reported that the few crude line features that exist, resolve into spots and streaks, not true regular continuous lines. They appear so with poor definition.

With the Mariner and Viking spacecraft pictures, the canals have disappeared into history. Nothing resembling them has been detected. There are what appear to be dried up river beds and large canyons but these cannot be identified for certain from Earth. Even so, some features appear as dusky streaks as can be seen on the drawings produced here. Perhaps the only area on Mars that has what can be described as having a Lowellian look about it is the region of Trivium Charontis and its surroundings. This is a small dark patch situated in the middle of a number of large bright regions and located at latitude +20°, longitude 195°. From it, many dusky streaks seem to radiate, adjoining it to the dark areas at north and south. It is intriguing to catch a glimpse of these features and reflect on the controversy they caused.

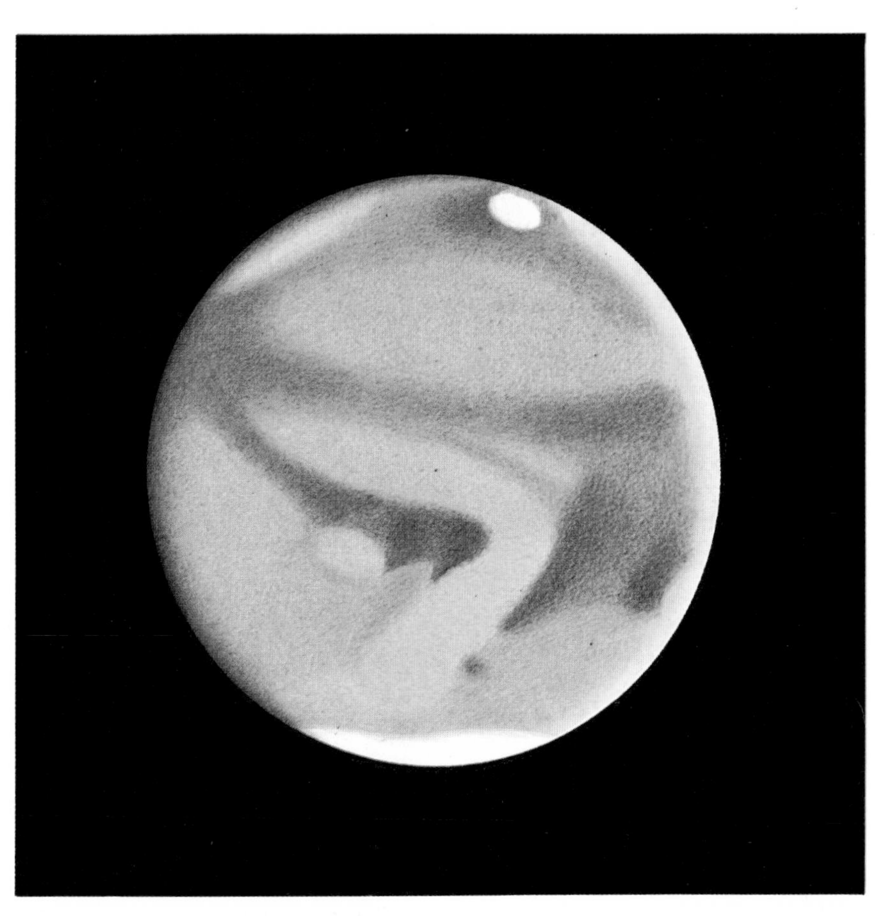

6.X.73 00.45 UT 254mm Refl ×400 Pandorae Fretum feature (above centre) is famous for its seasonal variations.

Surprising change

Mars can often spring surprises. Occasionally, well-known dark features may undergo considerable change in appearance—so considerable, in fact, that the area visible may be totally unrecognizable to the unsuspecting observer. In the main, the changes fall into three categories: those which are of a seasonal and regular variety; those which are seasonal but display certain irregularity; and those changes which are not really seasonal, but result in the appearance of completely new features. These may be a revival of a feature seen years before or may never have been recorded previously.

The seasonal changes were mentioned briefly while discussing the Martian canals. One of the most obvious is the appearance of a dark collar surrounding both the polar caps as they gradually shrink with the onset of the Martian summer. Another feature to show regular seasonal variation is Syrtis Major. Its preceding side, that is, the leading side as the planet rotates, will spread into the adjacent bright region, Libya, during the southern autumn and winter and reach its greatest extent shortly after aphelion. It will then appear much broader than it does during the southern spring and summer, being at its narrowest around the time of perihelion.

Following upon Syrtis Major as the planet rotates is a very prominent feature called Sinus Sabaeus, which is perhaps the most consistently dark marking on the planet. A lightish region known as Deucalionis Regio lies to the south of this area and still further south is another usually dusky region, Pandorae Fretum. This feature is subject to considerable change that is largely seasonal, but displays some irregularity. It usually appears dark and sometimes rivals Sinus Sabaeus in intensity during the southern summer, especially around perihelion. It will, however, normally appear faint or even invisible during southern autumn, winter and spring, though again, there may be some variation to this pattern.

Solis Planum

One area that experiences some seasonal variation but also undergoes considerable irregular change is the Solis Planum (originally Solis Lacus)—the most notorious of the variable regions. Normally the feature is somewhat oval in shape and may be very very dark. Sometimes, however, it may appear more extensive, consisting of numerous dark patches. So changeable is this feature that no set pattern can be established. Two drawings, given here, show the variation in its shape. One, made around aphelion, shows it to be a small, intensely dark round spot; the other, made near perihelion, shows it as extensive and patchy.

Changes that are not connected with the seasons are by no means uncommon. Perhaps one of the most famous was the dark streak across the Noachis region observed by E. M. Antoniadi in 1928. Some slight evidence of this was detected during the 1973 apparition and is indicated on one of the colour illustrations of that year, shown here. Certain major changes have occurred in recent years. A very extensive dusky region appeared on the normally bright region of Aethiopis during the apparition of 1958 and was again prominent throughout 1961. Though the region seemed to fade early in 1963 it again became obvious following the opposition of that year. But things had reverted to normal by the time of the apparition of 1965.

There has been a dramatic development in the regions known as Claritas and Daedalia, which are close to the Solis Planum. These had displayed their usual bright appearance during 1971, but in September 1973 a very dark marking was detected around longitude 110° latitude −30°. It was visible throughout the 1973 and 1975 apparitions and was often recorded as one of the darkest features on the whole planet. Past records show that there are certain similarities between the new feature and the development of a dark region in the same position during 1877. If, as it seems, this is a revival, it is strange that it should spend such long periods in obscurity. This type of change is of particular interest to the serious observer, since even in this day and age of space probes the exact reason for this phenomenon is not fully understood.

Seasonal variations

In the past it was generally believed that changes, particularly seasonal ones, indicated some sort of Martian vegetation since both the intensity and the actual colour of the regions changed. Lowell pointed out that the dark regions of the south appeared green before the southern summer and brown afterwards—just as we would expect vegetation here on Earth to do. It is not surprising that the vegetation theory was accepted for so long. On closer examination, these seasonal variations and changing colours are not as clear-cut as Lowell would have us believe. The soft land-

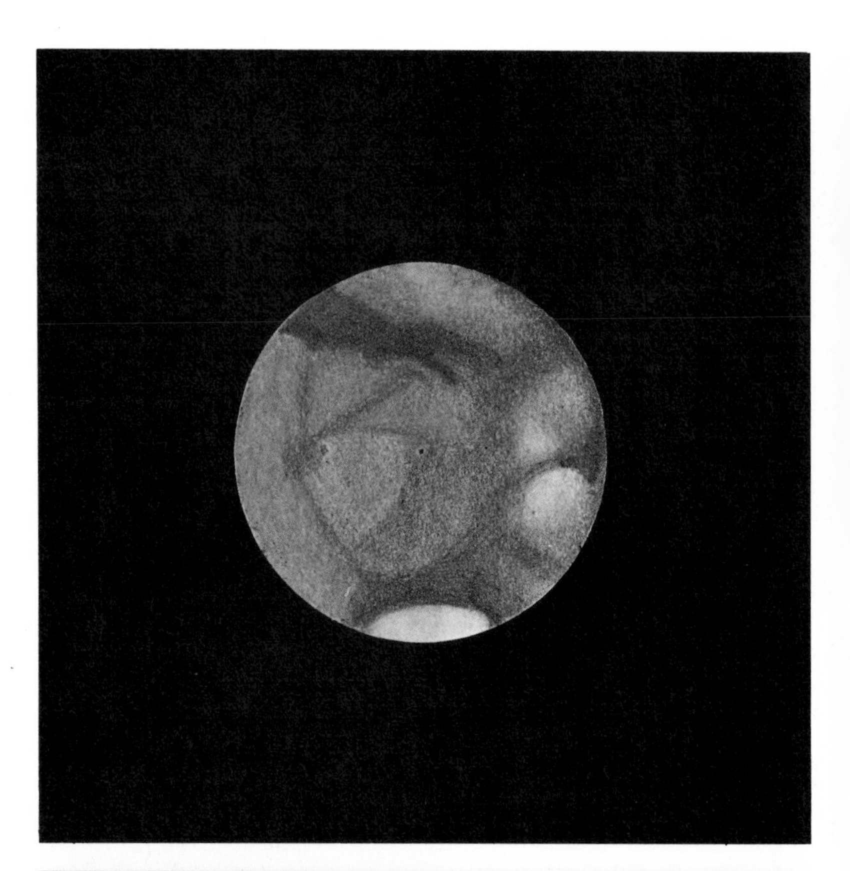

Above left *25.11.63 21.37 UT 225mm Refl ×274* Below left *15.III.65 23.45 UT 225mm Refl ×274 These show the change in appearance due to a temporary change or new feature. Dark shading of Aethiopis region has cleared two years later.* Centre top *1.IV.65 22.15 UT 225mm Refl ×274* Centre bottom *27.IX.73 00.30 UT 225mm Re ×274 Solis Planum is a regio irregular seasonal variation. In* centre top *the feature is experiencing winter showing*

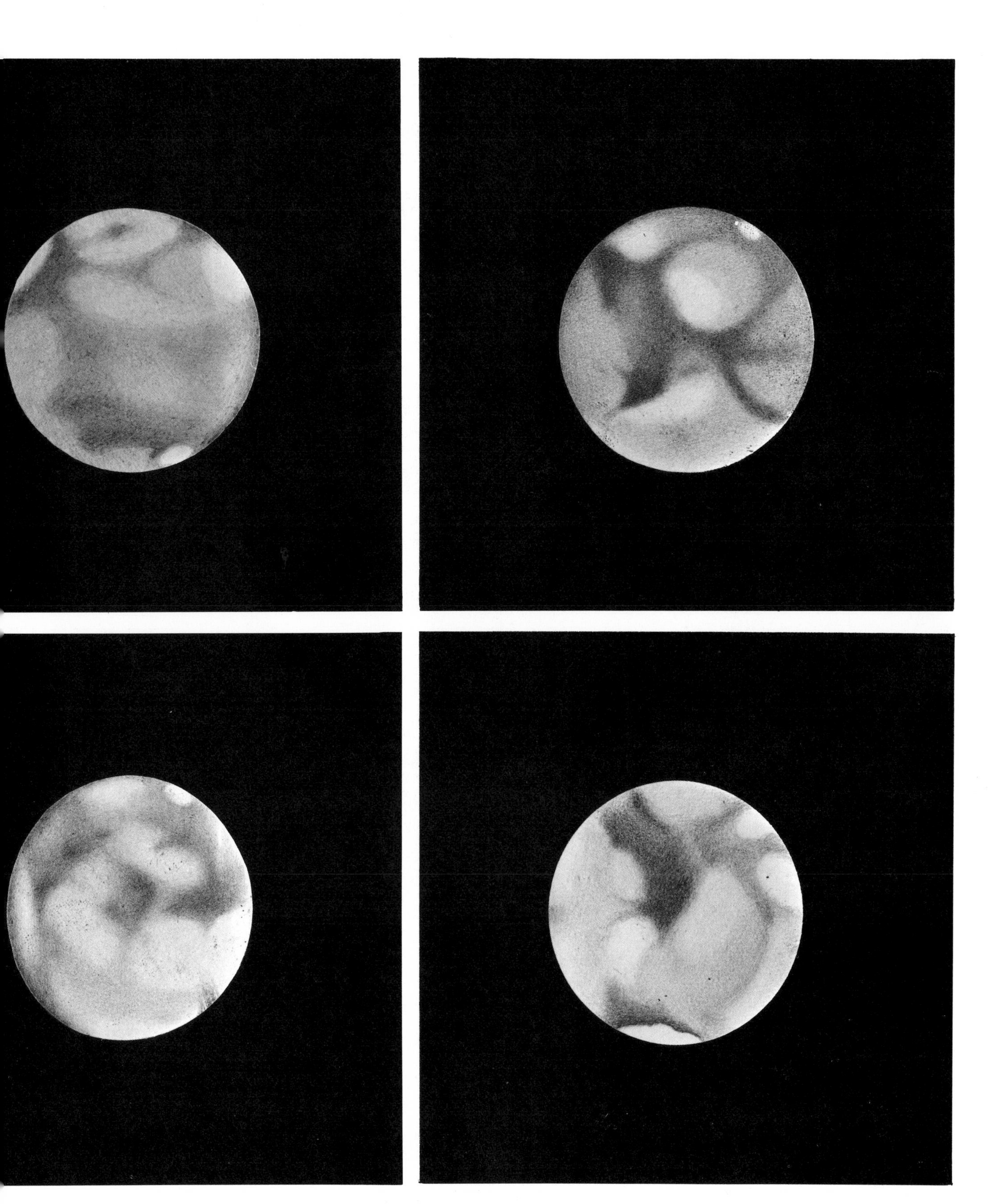

as a small dark spot. Below *it is central and experiencing summer. Note how extensive it is.* Right top *11.X.73 23.45 UT 254mm Refl ×380* Right bottom *26.I.78 21.40 UT 419mm Refl ×300 These two illustrations show clearly the seasonal variation in the Syrtis Major feature. The second illustration shows the feature as darker and broader than the first but features to the South East are less distinct, possibly due to haze in the Martian atmosphere.*

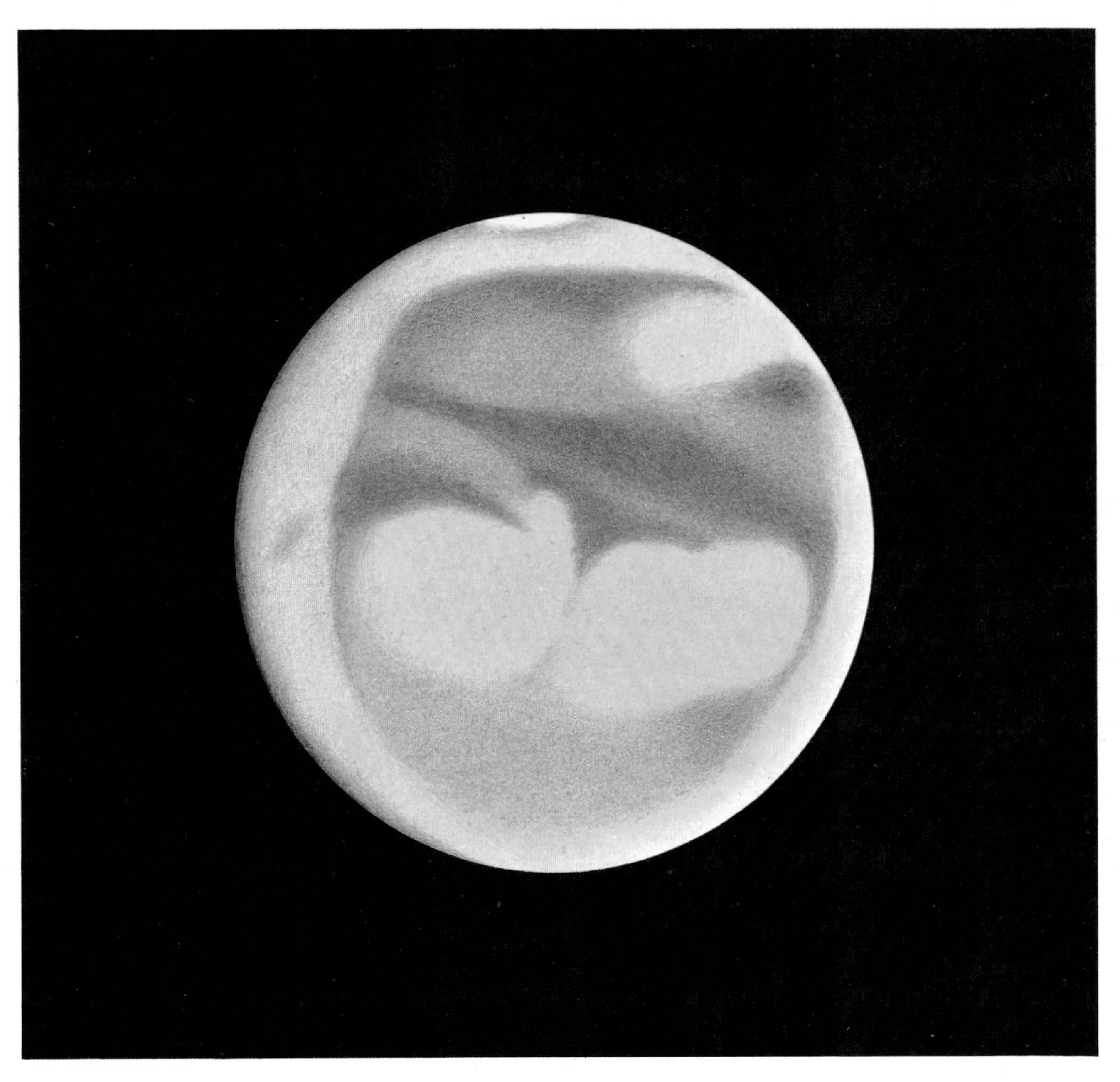

16.X.73 20.04 UT 254mm Refl ×250 Clear bright area, on the limb indicates a dust cloud.

ing spacecraft sent by the Americans have proved beyond doubt that there is no such plant life on Mars, although this fact was, in any case, fairly well established long before the spacecraft visited the planet. It is a pity but we must turn to other, probably less interesting, theories for an explanation of these puzzling changes.

One recent idea again made use of the melting of the polar caps. It was suggested that minerals in the Martian soil somehow become discoloured by a chemical reaction with either water from the polar melt itself or moisture in the atmosphere. Another, probably more feasible, theory once more relies on the Martian weather. The surface of Mars is mostly covered by a reddish dust, the colour of which is believed to be the result of oxidation of certain materials in the soil. Large amounts of dust are occasionally lifted into Mars' atmosphere by strong winds. (We know that at certain times in the Martian year winds with speeds in excess of 320 kph can occur.) As a result, there are extensive dust storms, which, on occasions, may be of such vast proportions that the dust is spread around the whole planet. After some time this dust settles and dark features often appear much less prominent than they did before the onset of the dust storm. Because of the reddish colour of the Martian soil, even a thin layer of this dust spread over the usually bluish-grey features will in all probability make them appear brownish and discoloured.

Dust storm

Lighter winds might in turn sweep these areas clear of dust and the features will appear to return to normal. Small-scale dust storms are particularly common and could be responsible for many of the localized changes reported from time to time. These changes may well appear

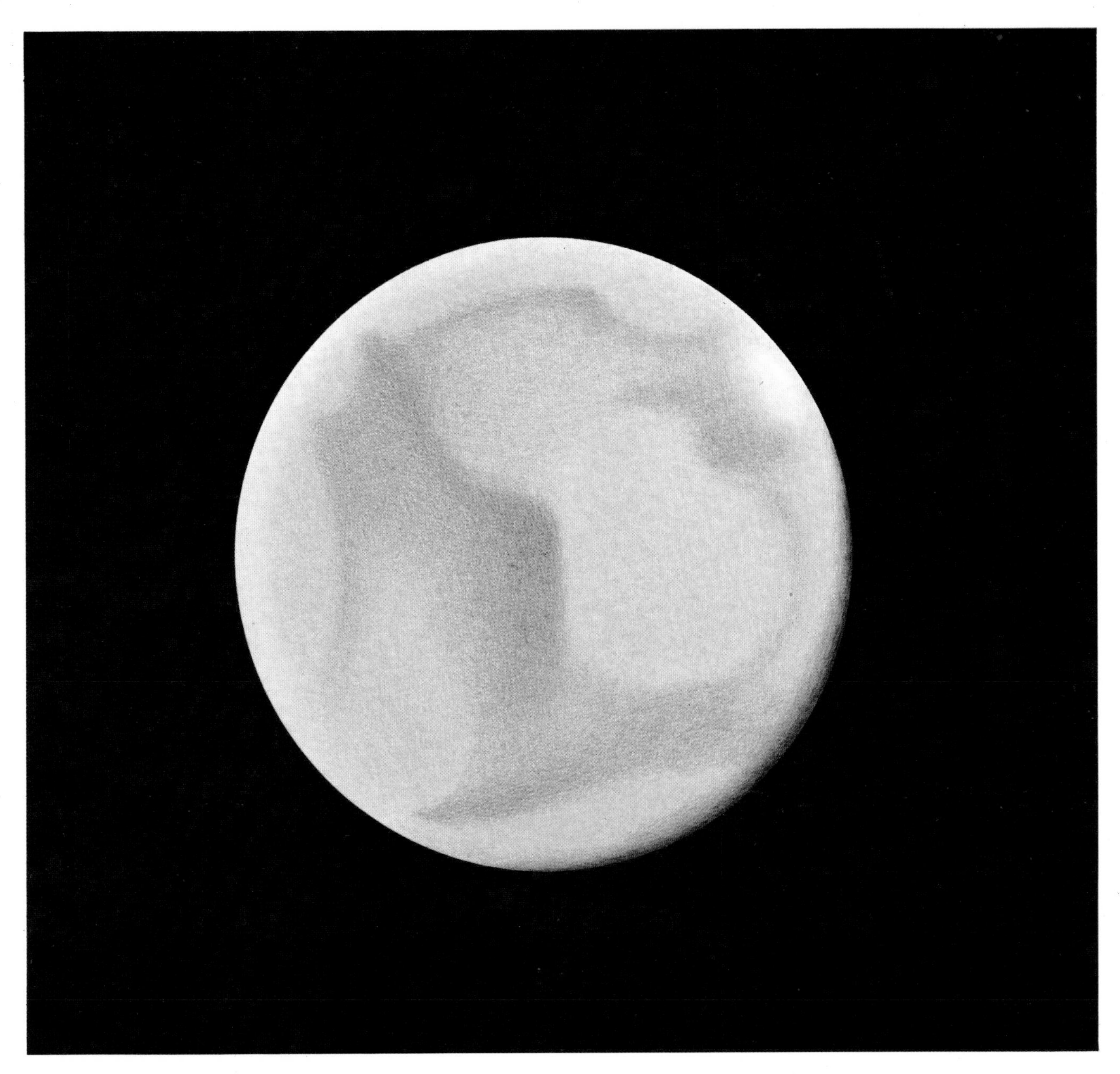

19.XI.73 21.30 UT 254mm Refl ×380 One month later the dark markings are still obscured.

seasonal since there are certain times during a Martian year when strong winds would be more likely to occur.

The atmospheric phenomena of Mars can be very dramatic and extremely interesting to watch. In fact, some of the most interesting have occurred in recent years. On September 21, 1971, a British amateur, A. W. Heath, detected a small yellow streak to the south of Syrtis Major. This streak signified the start of a great dust storm that was truly remarkable. The cloud spread rapidly so that by mid-October the whole planet appeared virtually featureless, with all the dark markings effectively obscured. It is worth mentioning that this storm began only three weeks after Mars passed the perihelion point of its orbit. The storm was still raging when the spacecraft Mariner 9 (designed to carry out an orbital reconnaissance of the planet) arrived in mid-November. It was so bad that at first no decent pictures could be taken and a considerable part of the Mariner 9 mission was affected. The dust eventually cleared, which the Mariner pictures showed in detail, but by the time of full clearance Mars was too far away for Earth-based study.

At the start of the 1973 apparition, the features of the Martian disc had reverted to normal, and stood out with remarkable clarity. This situation, however, was not to last. Observations up to the middle of October had shown no particularly unusual atmospheric phenomena, but between October 14 and 16 some clouds were observed around the Solis Planum area. On October 16 the first signs that another great dust storm was about to break out were detected. Observing at the time, the author discovered a large area of obscuration over the feature known as Mare Sirenum. In America, reports stated that considerable obscuration was evident in the Sinus Meridiani area and that it was rapidly spreading. So from this it appears that the storm had several points of outbreak.

It had all begun quite suddenly. By coincidence, October 16 was the night of Mars' closest approach to Earth and, more importantly, only a few weeks after Mars' perihelion. The large area of obscuration seen on the preceding, or sunset, limb that night appeared bright yellow, indicating dust. It was virtually clear of any dark shadings even though the region over which it was situated should have displayed one of the darkest features of the planet. Other parts of the disc remained clear but, on close inspection, more bright-yellowish areas could be seen over the south pole. Dust clouds, like yellow fingers, were creeping over the planet. As the rotation of the planet carried the main cloud out of sight other regions still seemed clear of dust. By October 21, however, virtually no recognizable dark markings could be seen. In a matter of five days, the dust had transformed the appearance of the whole planet. Despite good clear conditions here on Earth, Mars displayed only a blank, yellowish-pink disc. For the next three weeks virtually no visible change occurred in the state of the Martian atmosphere.

Return to normal

Although the storm was not as bad as that of 1971, only occasionally could the faintest signs of grey shadings be seen through the dust. Observations in early November suggested a slight, slow clearance of the dust and by the middle of this month many of the major features could be seen, though still with some difficulty, and only ghosts of their former selves. By the end of November, the clearance was well under way though all the dark features were well below their usual intensity and lacked colour. There was obviously a lot of haze remaining. The situation changed little during the period that Mars displayed a disc large enough to allow useful observation, which was at least until the end of January 1974. However, by the time that Mars had entered into the morning sky at the start of the 1975 apparition, things had returned to normal. Detailed observations of the disc showed no lasting effects of the storm. The area where the outbreak had apparently begun seemed unchanged after the dust storm.

Affected by the Sun

We have already remarked that both dust storms occurred shortly after Mars had passed through its perihelion point. This is interesting. A brief look into past records will show that this is not an uncommon occurrence. In 1956, which was another opposition occurring very close to perihelion, extensive dust storms became evident only a couple of weeks after perihelion. It is possible, though by no means certain, that these two things are connected. We may reasonably assume that when Mars is closest to the Sun its weather is in some way affected, to give rise to these vast storms. Mars is not always well placed for observation from Earth at the time of its perihelion and so some storms occurring would in any event be missed. It seems more than a coincidence that on the vast majority of favourably observed perihelia storms have been observed. There are many minor dust storms to be seen on the planet and this is something that the amateur should watch out for.

Dust storms, or yellow clouds, are not the only atmospheric phenomena easily observable with an Earth-based telescope. White clouds are frequently seen, at times so brilliant that they rival the polar caps. They show up best if a blue filter is used, unlike the yellow clouds described earlier, which are best seen in either yellow or red light. This fact suggests the different composition of the white clouds. Instead of dust they are composed of ice crystals and are generally situated high in the Martian atmosphere. A common appearance is a bright following or sunrise, limb, shown on most of the drawings illustrated here.

During the Martian night the temperature falls dramatically. As the night ends and the early morning Sun begins to warm things up, a morning mist, or fog, forms. This is seen as a bright limb, causing some obscuration of the features in that area, with normally bright regions appearing brighter still. The fog soon disappears with the rising Sun and is not carried any great distance into the sunlit hemisphere. On occasions a particularly prominent cloud may be seen, however, that will last through the Martian day.

One such cloud, discovered by E. C. Slipher in 1954, had a rather distinct W shape and was situated over the Tharsis–Amazonis region. This became visible during the Martian afternoon and remained until sunset. It is a recurrent phenomenon that has been recorded many times since its initial discovery. The cloud would seem to be situated over a particularly elevated region of the Martian surface.

At certain times of the Martian year a large area of cloud, or haze, may be seen over one or the other of the polar caps. This is usually most prominent at the end of winter for the hemisphere concerned. Drawings given here, made during the 1973 apparition, show the cloud, or polar hood as it is called, over the northern limb. What is shown is not the true cap, which if it were visible would not be so extensive—this is probably a result of the low-altitude Sun as seen from the region. The cap itself may at times be seen as a bright nucleus to the polar hood. Interestingly, though the northern cap suffers this cloud cover to a large extent, it is never so great as the cloud cover that may affect the southern cap. This is a direct result of the fact that southern winters occur when Mars is at its greatest distance from the Sun.

Polar caps

The actual waxing and waning of the polar caps themselves is instructive to watch. A brief description of what can generally be expected of the southern cap will make this clear. At the end of the southern winter the cap is hidden almost completely by cloud cover, as described above. Through spring the cap shrinks, and at an ever-increasing rate as the summer approaches. At first, the melting of the cap is hidden by the cloud cover, but this disappears before summer. When first uncovered the cap is brilliant white, but as melting progresses certain dusky features start to show through and the cap loses its stark whiteness. It may divide into areas of various sizes and brilliance, but one particular area, in the region south of the feature Hellas, gradually becomes detached from the main cap and eventually forms an isolated white spot long after the cap has receded from the area. The patch itself finally begins to break up and disappear around mid-summer. This detached portion indicates particularly high ground and the region has been given the name 'Mountains of Mitchel'.

During high summer the cap may seem to disappear without trace, though it may not have actually done so. The centre of the southern cap is situated about 7° from the true pole, and, if it is small enough, may at times be hidden from view. Yet, twelve hours later it would be brought into view by the rotation of the planet.

Deimos and Phobos

The cap appears at its smallest at the end of summer. During early autumn patches of cloud become visible in the area and eventually cover the cap. The tilt of the planet takes the cap from view and it remains out of sight throughout the late summer and winter.

The northern cap undergoes a similar series of changes but, whereas the maximum extent of the southern cap may be 60°, the northern cap is usually never more than 50°. The northern cap never becomes as small as the southern cap, either. The reason for this greater change in the south cap is that the southern summer occurs when Mars is at perihelion and the winter when at aphelion. The opposite is the case for northern summer and winter, so the variation of the north cap is to some extent evened out.

Mars, unlike the two inner planets, has the attendance of two natural satellites, or moons. These are tiny and much smaller than our Moon. Phobos, the larger and closer of the two, has a diameter of 23 kilometres and is 9,400 km from the centre of Mars. Deimos is only 13 km in diameter and orbits at a distance of 23,500 km from the centre of Mars. Their orbital periods are 7 h 39 m and 30 h 18 m respectively. The orbital period of Deimos results in this satellite remaining visible to a point on the surface of Mars near to the equator for three days without setting. During this time it would appear to go through the full range of phases twice over, but would remain small and faint. The motions of Phobos would be stranger still. Its distance from the centre of Mars means that it is only a little over 6,000 km from the surface.

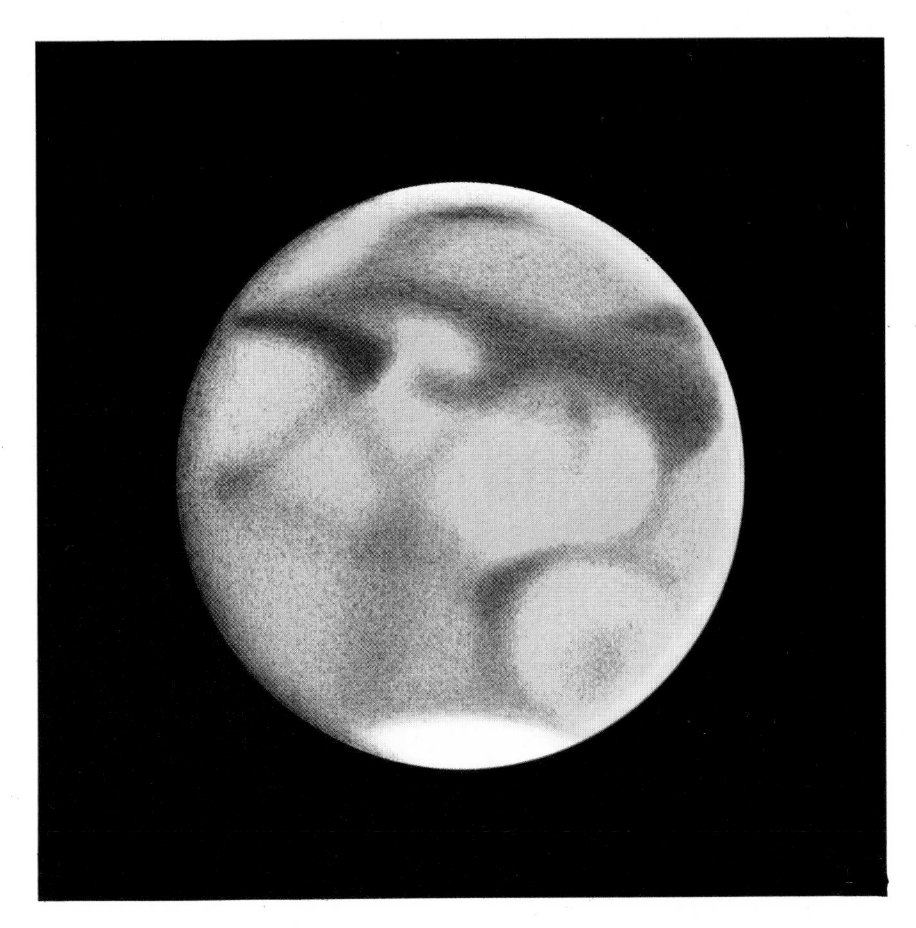

13.XII.75 21.30 UT 254mm Refl ×350. Features of Mars restored after the dust storm of 1973.

Phobos is the only satellite known that orbits the primary in less time than the primary takes to turn once on its axis. As a result, to an observer on Mars it would appear to rise in the west and set in the east and be above the horizon for four hours at the most.

Mariner 4

Earth-based observations of these two satellites is far from easy. Their discovery in 1877 was made by Asaph Hall, using the 650-mm Washington refractor. It needs a reasonable-sized telescope to pick them out. Their opposition magnitudes are +11.6 for Phobos and +12.8 for Deimos. The glare from the planet makes observation all the more difficult and it is reckoned that a telescope of at least 200-mm aperture is required to see them. When at opposition, their apparent distances from the centre of the primary are $1\frac{1}{2}$ and 3 primary diameters respectively. To aid with their identification an occulting bar, described in the Venus chapter, should be used. Though it may be of interest to spot these bodies, there is nothing the amateur can do to add to our knowledge of them.

With the exception of the Earth and the Moon, Mars has been more closely studied by robot spacecraft than any other planet. Most of the success has been on the American side, and we owe a great deal of what we know about the planet to them. It is worth summarizing the various Mars probes to see how our knowledge of the planet has built up.

The first big shock came in July 1965 when the Mariner 4 spacecraft passed within 9,850 km of Mars' surface, sending back to Earth the very first close-up pictures of another planet. Mars had for ages been considered as the planet most resembling the Earth and, although observations were beginning to indicate that this was not true, few people really expected the apparently lunar-type landscape revealed by these first pictures. Twenty-two photographs were

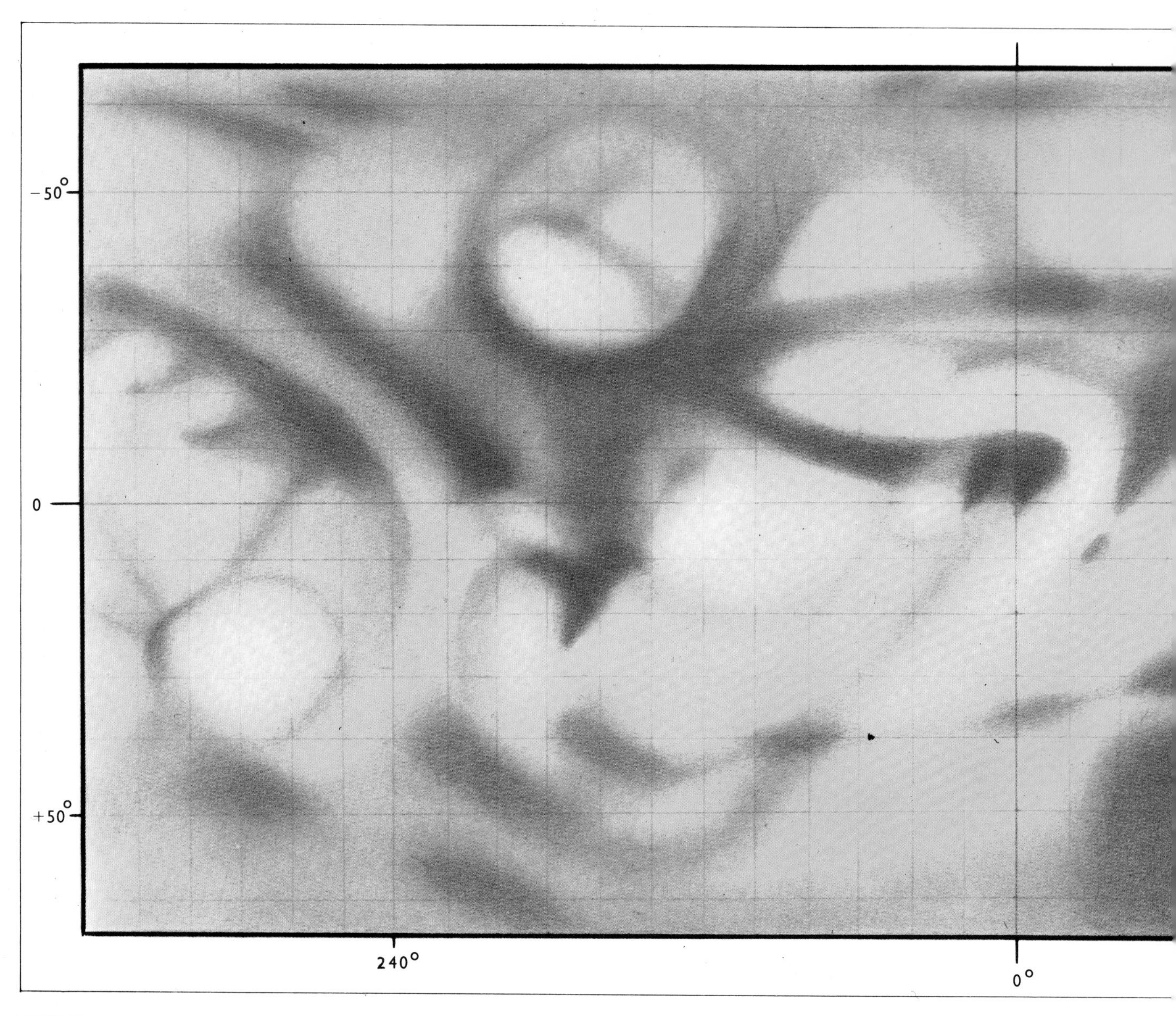

−50°
0
+50°
240°
0°

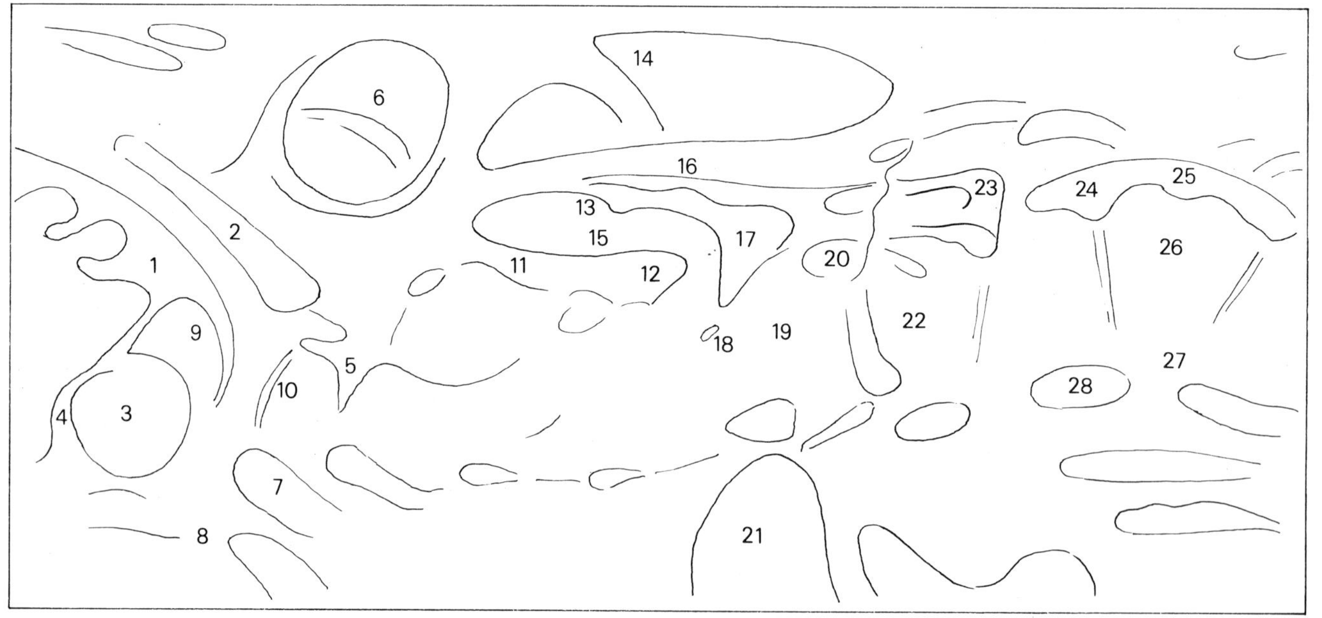

1
2
3
4
5
6
7
8
9
10
11
12
13
14
15
16
17
18
19
20
21
22
23
24
25
26
27
28

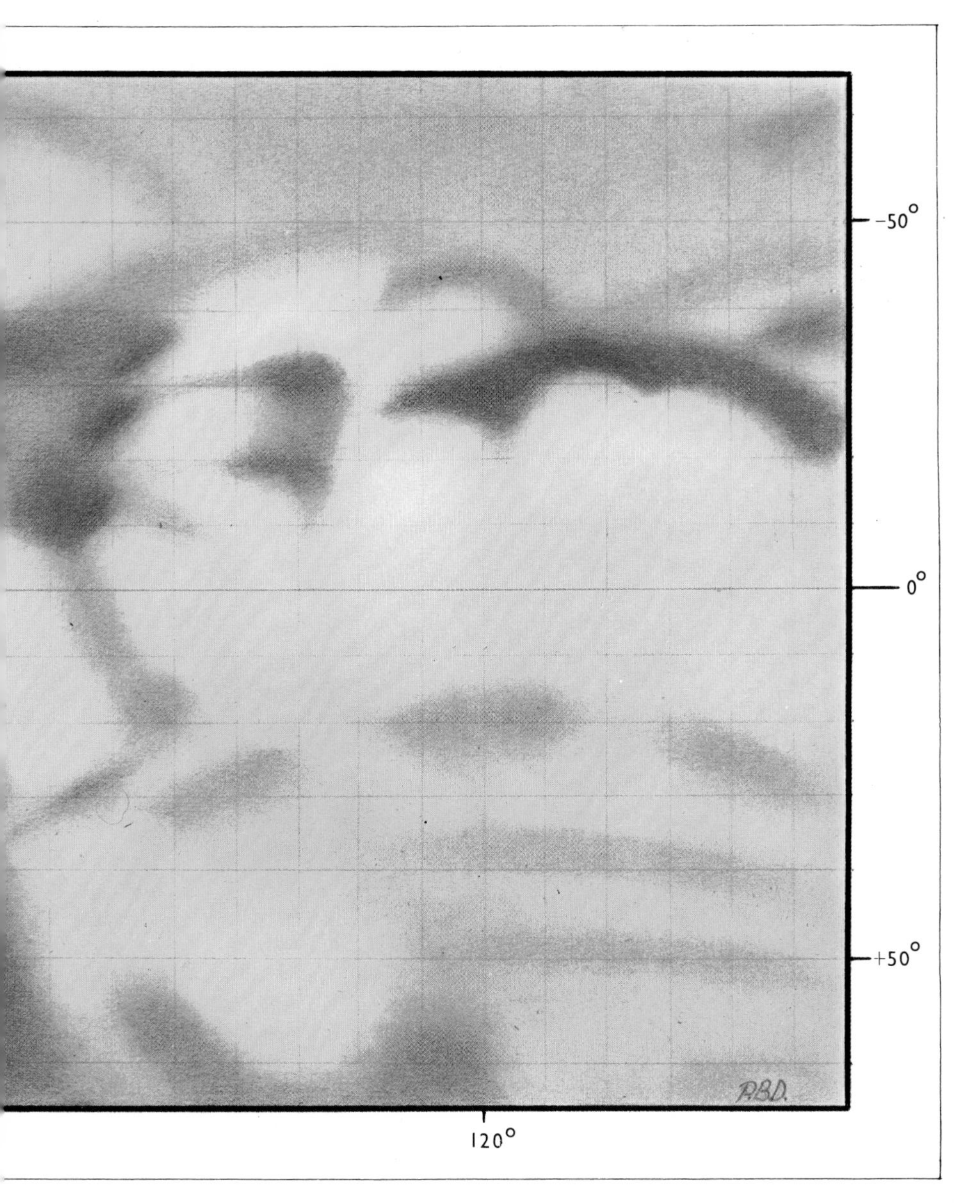

Author's map of Mars from observations of 1963–1975.

1 Mare Cimmerium
2 Tyrrhena Planum
3 Elysium
4 Trivium Charontis
5 Syrtis Major
6 Hellas
7 Casius
8 Utopia
9 Aethiopis
10 Thoth Nepenthes
11 Sinus Sabaeus
12 Sinus Meridiani
13 Pandorae Fretum
14 Noachis
15 Deucalionis Regio
16 Mare Erythraeum
17 Sinus Margaritifer
18 Oxia Palus
19 Chryse
20 Aurorae Planum
21 Acidalia Planitia
22 Tharsis
23 Solis Planum
24 Claritas-Daedalia
25 Mare Sirenum
26 Memnonia
27 Amazonis Planitia
28 Olympus Mons and Arsia Mons

returned to Earth, and these covered just a small portion of the Martian surface in the general region of Amazonis and Mare Sirenum. Also, resolution was quite poor by today's standards. Nevertheless, the pictures sufficed to show a large number of craters, 70 in all, the biggest of which had a diameter of 112 kilometres. The craters, on first inspection, appeared Moonlike but when examined they were found to be much shallower than the average lunar crater, and the slopes of their walls seemed much more gentle.

Martian atmosphere

Another interesting result from the Mariner 4 mission was the measurement it gave for the atmospheric pressure at Mars' surface. This was arrived at in a very ingenious way. Owing to the path of Mariner 4 it would at one time pass behind Mars, or into occultation, as viewed from the Earth. Although all contact would be lost while the actual occultation took place, there would be a short time, both prior to, and following, occultation when the spacecraft passed behind the atmospheric layer of Mars. By sending out radio signals of a known frequency to the spacecraft and having these transmitted back during the period that the craft passed behind the atmosphere, any change in the frequency returned, as compared with that sent out, would be a measurement of the planet's atmosphere. In this way it was found that the surface pressure was only 1% of the Earth's at sea level. This was $2\frac{1}{2}$ times less than the generally accepted value prior to the probe. Temperature measures were very low at −113°C, but the region measured was quite close to the southern polar region which was then experiencing its winter. The main constituent of the atmosphere was found to be carbon dioxide and it was estimated that, at the poles during the Martian nights, temperatures would be low enough to freeze some of this carbon dioxide out of the atmosphere, forming a thin deposit of dry ice on the surface. From this it was assumed that the polar caps were composed of frozen carbon dioxide (dry ice).

Although the craters came as a shock to most people, there have been at least two astronomers in the past who have recorded such features by telescopic observation. J. Mellish, using the 1-metre refractor at Yerkes Observatory, and E. E. Barnard, using the lick 900-mm refractor, made observations of these features, but little is known of them. In Barnard's case it is thought that he was afraid to publish for fear of ridicule. The remarkable thing is that the latter's observations were made as long ago as 1892–93. If nothing else, this indicates just what can be achieved by visual observation, given the right instrument and conditions. It is a pity that their results were not published and accepted. What a tremendous triumph the discovery of craters on Mars would have been for visual observation and for astronomy!

Successful probes

The next successful probes were Mariners 6 and 7, which passed Mars in 1969. Mariner 6 passed to within 3,430 kilometres of Mars' surface on July 31 and Mariner 7 to within 3,200 kilometres on August 5. Although like Mariner 4 these were only fly-by missions, the clarity of the photographs was far superior and told us a great deal more about the Martian surface feature. Mariner 6 followed a path which allowed it to study the equatorial regions of the planet. A little later, Mariner 7 followed a path that enabled study of the south polar region, taking 126 photographs altogether, almost three-quarters of which were far encounter pictures.

Most of the regions shown were covered with craters of all sizes, which for the most part seemed fairly shallow. Craters near the polar caps were shown to have considerable frost on their rims and those within the polar cap itself were also photographed. Many craters that skirted the edge of the ice cap were shown to have conspicuously dark floors. Lots of hills and grooves were detected but, strangest of all, was the feature Hellas. This was virtually devoid of any detail, its whole expanse appearing clear. Mid-day temperatures at the equator were found to be a pleasant 15°C, but night temperatures were in the region of −70°C. The temperatures at the poles were, of course, much lower and the theory that the ice caps consisted of frozen carbon dioxide seemed to be substantiated.

On November 14, 1971, one of the most highly successful planetary probes ever entered into orbit around Mars, and Mariner 9 thus became the first artificial satellite of another planet. As the probe closed in on the planet the far encounter pictures showed a virtually blank disc. It will be remembered, from earlier in this chapter, that at this time the great 1971 dust storm was in full swing. Unfortunately, when the probe went into orbit and began to send us its first close-range photographs, all that could be seen was the upper layers of dense dust clouds. Through the dust, a few tantalizing glimpses of odd features were obtained, but it was nearly three months after the probe's arrival that the first really clear pictures began to come through. At that time they realized that the few spots that had been seen were the summits of incredibly high volcanic formations, one of which was the feature shown on Earth-based maps as Nix Olympica and possibly the most impressive of all formations on the surface of the planet. Across its base it measures at least 500 kilometres, and in height an imposing 24 kilometres, with a caldera 65 kilometres in diameter at its summit. We have nothing comparable to this on Earth—even our highest mountain is only just over one-third as high. The feature now goes under the name of Olympus Mons, or Mount Olympus.

Splendid pictures

Apart from these huge volcanoes, which seem to be inactive at present, there were many other magnificent formations, scarcely hinted at by previous probes. This is not really surprising as the photographs sent by the first three Mariners only covered 20% of the total surface. Mariner 9 was, of course, giving us 100% coverage and at all angles of illumination. A further fine discovery was that of a huge canyon 4,000 kilometres long by as much as 120 kilometres wide in places and 6 kilometres at its deepest. Compared to this, our own Grand Canyon would be a minor tributary. Other features of interest were what appeared to be numerous dried-up river beds and extensive regions covered with sand dunes. Pictures of the Hellas basin showed a little detail and some localized dust storms. This basin had possibly been smoothed by the collection of dust, but from its brightness variations may well be the source of many dust storms.

Excellent coverage of the melting of the south polar cap was obtained during the Mariner 9 mission and as the ice receded a strange sort of stratification was revealed near the pole, now referred to as lamination. The fact that there is virtually no crater formations in this region suggests that this layering is a fairly recent development. For most of the mission the northern cap was covered with cloud but, towards the last few weeks, the cap itself was revealed and as it receded showed similar layering in the terrain.

Viking

The first close-up pictures of Mars' tiny moons came from Mariner 9 and, of the two, Phobos showed more clearly as an irregular, slightly elongated body with numerous crater pits, some of which are quite large, considering the size of the moon itself. In almost one year, Mariner 9 transmitted over 7,300 pictures, providing more information than we had gathered in the whole history of observing the planet. Transmissions finally ceased on October 27, 1972, but the success of the Mariner 9 mission had even exceeded the hopes of the planners.

In 1976, two more probes went into orbit around Mars, Viking 1 on June 19 and Viking 2 on August 7. Thus began a programme of exploration that was to make even Mariner 9 seem ordinary by comparison. The difference here was that both spacecraft comprised two vehicles. Each has an orbiter, to carry out a more detailed orbital reconnaissance, and still more important, each had a lander, which would actually sample the Martian soil and search for life.

When the first Viking spacecraft settled into orbit, even more interesting features were photographed. Islands, formed by some erosive force, surrounded certain craters while unusual flow patterns surrounded others. These flow patterns give the distinct impression that whatever caused the craters also heated up the surrounding surface, which possibly then became molten. As this spread around the craters it rapidly cooled to become solid again, thus freezing the shape of the flow. It has even been suggested that beneath the Martian soil there might be a considerable amount of water ice locked in the soil as permafrost. If this were so then the heat generated by the formation of a crater would momentarily melt the permafrost, causing sludge to flow away from the crater in all directions. It would, however, freeze again almost immediately and the flow would be stopped in its tracks. This, too, would give rise to the unusual effect seen by Viking.

Despite all the wonderful formations that were being revealed, the most important task for the orbiter at this stage was to examine the landing site, which had been chosen from Mariner 9 photographs. This was a smooth plain in the region called Chryse Planitia, a site where, it was felt, the chances of finding life were high. The improved resolution showed the area to be too rough and another site had to be selected. Finally, a region in the same general area was chosen.

Landing on Mars

On July 20, 1976, 3½ hours after leaving the orbiter, the Viking 1 lander touched down safely on the surface of Mars. During the final descent studies of the atmosphere showed the presence of about 3% nitrogen, which heightened the hopes of finding life. Once on the surface reports of weather conditions were received: winds were at most 24 kph, easterly in the afternoon and southwesterly after midnight; atmospheric pressure was 1% of Earth's at sea level, confirming the Mariner findings, and temperatures ranged from —86°C shortly after dawn to —31°C at mid-day. The atmosphere consisted of 95% carbon dioxide, 3% nitrogen, 1½% argon and a small amount of oxygen.

Almost as soon as the lander reached the surface, the first pictures began to come through. The surprising thing was the brightness of the sky. It had been expected to some extent but, despite the rarity of the Martian air, it looked almost as bright as our own sky. Its colour was distinct salmon pink, which undoubtedly resulted from the reddish dust suspended in the atmosphere above the planet.

The surface was a mass of small boulders and numerous small dunes of fine-grained soil—just like some desert scenes here on Earth. The colour of the soil was very red, thought to be due to a limonite coating. Some rocks, however, did have a slight greenish cast. Next, after some days on the surface, an actual sample of soil was taken by means of a small mechanical grab at the end of an extending arm. Three different tests were then carried out on the soil in the hope that the presence of living organisms were to be found.

The first test involved placing a small soil sample into a sealed chamber. Carbon dioxide and carbon monoxide were pumped in, the carbon atoms being radioactive in order to enable them to be traced later. The soil was then subjected to simulated sunlight. Any living organism should have turned the carbon gas into organic compounds. After a suitable incubation period the chamber was flushed clear of the gas and the soil heated to vaporize any organic compound. If these had been produced during the incubation period, then radioactivity would be detected. The experiment was, however, inconclusive.

A second experiment involved the addition of a liquid nutrient to the soil sample and, once again, the carbon atoms in the nutrient were radioactive. Any organism feeding on this would release gases into the atmosphere of the chamber in which the sample was contained. If radioactivity were detected in this atmosphere, this would indicate the presence of living organisms. Again, there was no positive result. Radioactivity was detected but things did not act as they should and the question remained open.

The third experiment took a soil sample and soaked it in a very rich liquid nutrient. Any organisms within the soil would release gases and bring about a change in the atmosphere of the chamber containing the sample. Although at first this gave a promising result, the process did not continue as expected and the experiment must once more be regarded as inconclusive.

While this was going on, the Viking 1 orbiter was still sending back remarkable pictures of Martian surface features, such as the beautiful colour pictures of the great Valles Marineris. Other canyons on higher ground were shown full of bright mist or fog as they began to warm up in the early morning sun. Huge landslides were depicted on the floors of some canyons with great lava flows over large areas of the surface. Streaks indicating wind action showed clearly on many parts of the surface. Fields of dunes, some 40 kilometres long, were revealed in great detail and collapse features were evident in many places.

As if all this were not enough, the Viking 2 mission was equally successful. On September 3, 1976, its lander touched down also in an area covered with rocks. The landing area this time was in the region Utopia Planitia. The rocks at this site were full of small holes, or vesicular, suggestive of a volcanic origin. Not many sand dunes were visible at this site, but there was what appeared to be a dried-up river bed quite close to the lander. It was not certain whether this was the result of water flowing or of wind erosion. The experiments to detect life met once more with inconclusive results, but some of the Viking 2 orbiter photos were magnificent. The sedimentary layers in the surface at the poles were photographed in very great detail and on some frames great overhangs of ice on large scarps can be seen.

Each of the Martian moons was also photographed in closer detail than ever before, enabling vastly superior maps to be compiled. Phobos exhibits some remarkable features. It is only 27 kilometres long yet it has one crater with a diameter of 10 kilometres. This, incidentally, has been given the name Stickney, after the wife of the discoverer of Mars' two moons. She is said to have greatly encouraged the search that resulted in their discovery. Phobos is also marked with numerous parallel grooves—each of which has the appearance of a crater chain—averaging 150 metres in width and up to 20 metres in depth.

Large-scale changes

One theory as to how they were formed suggests that they result in turn from the formation of the crater Stickney, from where they appear to radiate. This is very probably true. Tidal stress is another possibility worth considering. Deimos, like Phobos, is irregular in shape and there are also numerous craters on its surface, making it just as heavily cratered as Phobos. The craters are, however, shallower and the general appearance of Deimos smoother.

The vast store of information re-

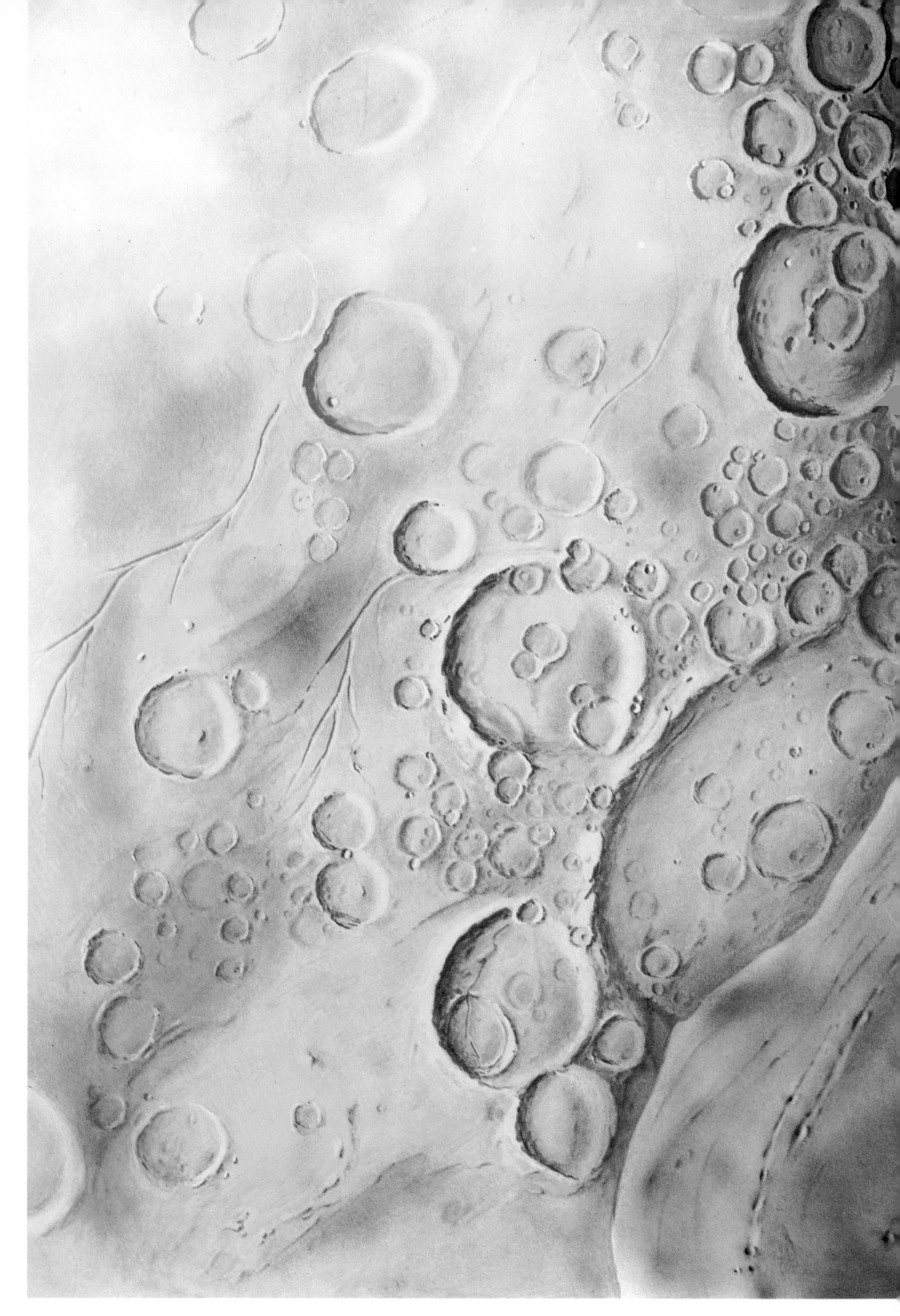

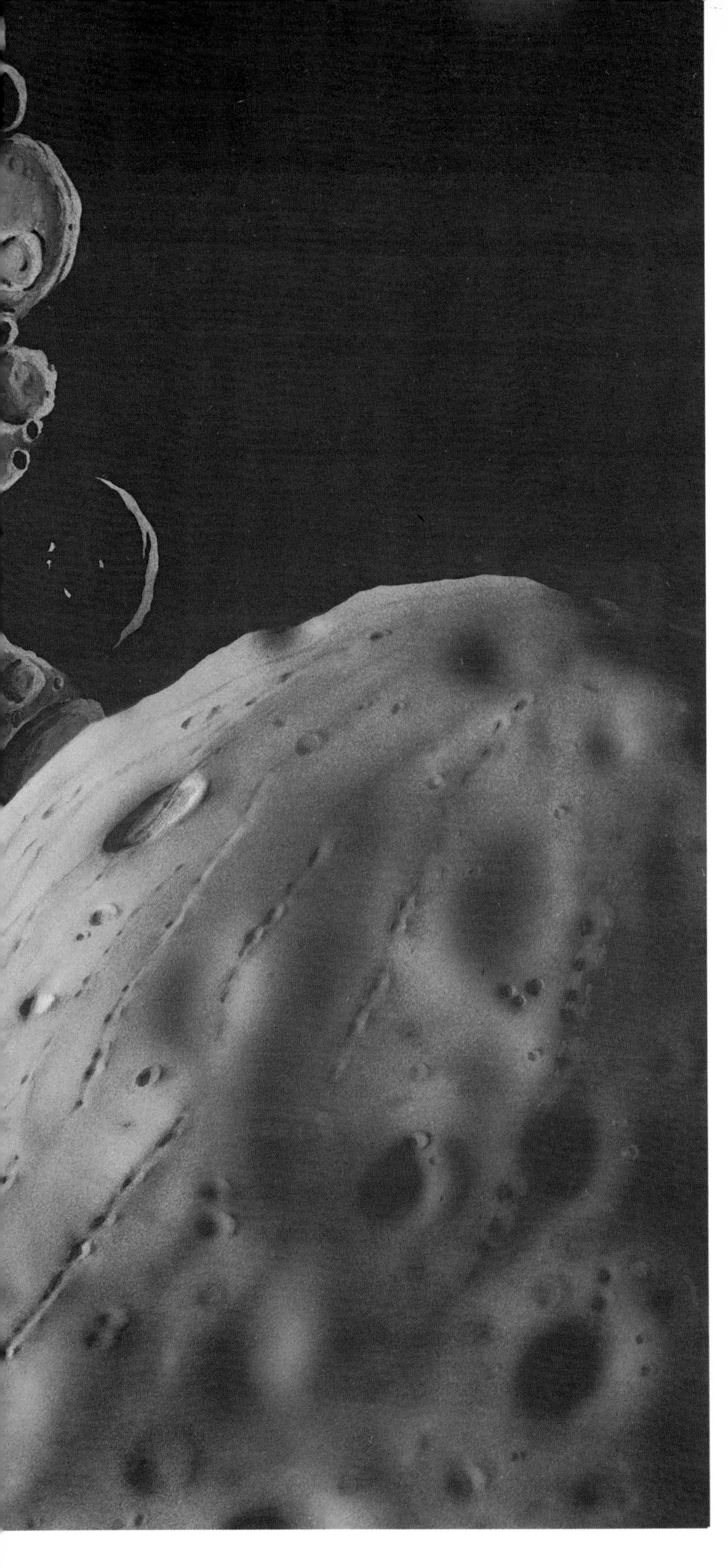

ceived from the Mariner and Viking spacecrafts will take years to evaluate properly. In the meantime we are at least left with a very good idea of what the planet looks like and why it appears as it does.

Scientists are now fairly certain that some large-scale changes have occurred on Mars in the recent past, that is, within the last few tens of millions of years. It seems that water once flowed in abundance and took a hand in shaping the surface as we see it. Perhaps this water is still there, frozen beneath the surface. The general consensus is that the polar caps are largely water ice, the dry ice theory having fallen out of favour. Many of the spectacular features suggest flowing water, which may be brought about by localized heating from small-scale volcanic activity. It has even been suggested that the spectacular Olympus Mons might be a huge mountain of ice—a theory borne out by the fact that such a large build-up of pure rock is difficult to explain whereas a build-up of ice is quite acceptable.

One theory put forward by Carl Sagan suggests that if the axial tilt of Mars were to increase it might be possible that both polar caps would vaporize completely, releasing volatiles that would increase the density of the planet's atmosphere. The result would most certainly be a change in climate.

All of these theories are, of course, just ideas and the field is wide open to such speculation. One thing does seem to emerge from all of this for sure, however, and that is the fact that the appearance Mars is presenting at the moment is only temporary. If we could see the planet in a few million years' time we might not even recognize it. Who knows? Perhaps continued observation with Earth-based telescopes will help to answer some of the outstanding questions—not despite the space probes but because of them. A view of Mars with a moderate-sized telescope is still a sight to behold, putting one in mind of a brightly coloured rubber ball, and all the highly detailed photographs now available can take nothing away from it.

Mars as it might appear from Phobos. Man may one day stand on Mars's closer moon.

Minor Planets

After Mars, the next planet in order of distance from the Sun is Jupiter. Whereas all the planets previously discussed are spaced at fairly regular intervals, with distances between each orbit that are not too great, once outside the orbit of Mars we are suddenly confronted with an enormous gap. In fact, we have to travel a distance almost three and a half times that of Mars from the Sun before we encounter the orbit of Jupiter. This is an odd situation and it seemed so to earlier astronomers. It was almost as though another planet should be orbiting the Sun somewhere in this space. A simple numerical law, known as Bode's law, was even devised, that agreed nicely with the known distances of the planets when expressed in astronomical units (one astronomical unit being equal to the mean distance of the Earth from the Sun, or 149.6 million kilometres). However, this law gave one value, corresponding to a distance of 2.8 astronomical units from the Sun, a position between the orbits of Mars and Jupiter, and one where no planet existed. So weighty was the evidence considered that, at the end of the 18th Century, a group of astronomers headed by J. Schröter, got together to undertake the search for such a body, in case it existed.

First discovery

Unfortunately, before they could put their search into full operation, an Italian astronomer, Father Piazzi, while in the course of charting the stars, noticed one which appeared to move slowly with respect to other stars in its vicinity. Only members of the Solar System do this and so, on the memorable night of January 1, 1801, a new addition to the Sun's family was discovered. Its distance from the Sun was only slightly under 2.8 astronomical units, or 414.4 million kilometres. This satisfied the numerical law by placing a body between the orbit of Mars and Jupiter. Astronomers were not completely satisfied, however, because of a slight problem, namely, that the body was literally minor. Its diameter was originally estimated at 770 kilometres,

Small rugged body of an earth-grazing asteroid passing within a few million miles of Earth.

though more recently its diameter has been evaluated at 1,003 kilometres. Even so it is tiny when compared with the major planets. It was, however, an important body and thus worthy of a name, so it was duly christened Ceres, goddess and guardian of Sicily, the country from which its discovery took place.

Similar orbit

This is not the end of the story, it is only the beginning. The group of astronomers continued their search, possibly feeling that Ceres did not really fit the bill, and in March 1802 H. Olbers, a member of the group, discovered a second small body with an orbit virtually identical to that of Ceres. This object was given the name Pallas and was found to be even smaller than Ceres. Recent estimates give its diameter as only 608 kilometres.

Following these discoveries, theories were advanced that perhaps these bodies were the fragments of a larger body, which for some reason had broken apart. If this were so, then it was reasonable to assume that other such objects might be found. Sure enough, in 1804 a third small body was discovered, whose orbit resembled very closely those of the other two. The name of Juno was selected and its diameter was again even smaller, at 247 kilometres. This time, another member of the group, K. Harding, was the discoverer. Finally, in 1807 Olbers discovered a fourth minor planet, which now goes under the name of Vesta. Its orbit was again similar to the previous three, but its diameter was close to that of Pallas, at 538 kilometres.

The group of astronomers broke up around 1815 and it was some time before any further discoveries were made. The fifth minor planet, Astraea, was not found until 1845. It too had a similar orbit and a smaller diameter still, only 117 kilometres. This fifth discovery sparked off a positive rash of finds and the total had reached an amazing 322 by 1891. In that year, No. 323 was discovered by Max Wolf, using the then new photographic method. He himself added more than 100 others to the list, which has since grown at an ever-increasing rate. To date the number of discoveries is well over 2,000 and is still on the increase.

The first four of the minor planets are easily seen, even with small telescopes. Vesta is the brightest of all the minor planets, as viewed from Earth, and is the only one that can at times attain naked-eye visibility. The maximum magnitude is around +5.8, which is for an opposition occurring when Vesta is at perihelion. Normally the magnitude is less, though even at unfavourable oppositions it is never much below 7th magnitude. Binoculars will pick it out with ease. Telescopically it looks like a yellowish star, as in fact do all the minor planets, hence the other name by which they are known, 'asteroid'. The actual angular diameter of Vesta can at times be 0.6″, or just over half a second of arc. It is therefore possible in exceptional seeing conditions to pick out its tiny disc with telescopes of a moderate aperture.

The maximum opposition magnitudes of Ceres and Pallas are +7.4 and +7.8 respectively, and so these too can, at favourable oppositions, be easily picked up with binoculars. Juno, number three of the first four discovered, has a maximum magnitude of only +9.2, and a small telescope is really needed to see this favourably. There are some asteroids that attain magnitudes comparable with this and dozens, in fact, with maximum opposition magnitudes greater than +11.

Icarus

However interesting it may be to locate these objects, once this has been done there is not, in general, a great deal left to do other than to watch their motion among the stars from night to night. But there are a few exceptions. As already described, most of the asteroids have orbits confined to the region between Mars and Jupiter and, as a result, their distance from Earth is usually considerable. A few have highly eliptical orbits, which take them out of this region and bring them, at their perihelion, closer to the Sun than to the Earth.

The asteroid Icarus has an orbit with an aphelion distance of nearly 300 million kilometres, some way beyond the orbit of Mars, yet at perihelion it approaches to within 32 million kilometres of the Sun, closer even than Mercury. It is believed that at such times it might glow red hot, which is how it got its name. The orbit of Icarus will at times cross the Earth's orbit, and if the Earth happens to be in the vicinity when this happens a close approach will result. At best, Icarus can come to within six and a half million kilometres of the Earth, which is roughly 16 times our distance from the Moon, but since it is nothing more than a huge lump of rock, 1 kilometre across at its greatest, it never becomes a conspicuous object. With a maximum magnitude of only +13, a moderate-sized telescope is needed to show it well. Such close approaches are quite rare and should, if possible, be followed carefully. The motion among the stars can be fairly rapid and should be easily detected after only a couple of minutes.

'Earth grazers'

Some asteroids can come much closer to the Earth than Icarus. One of these 'Earth grazers' passed us at less than 2 million kilometres away, little more than five times the distance of the Moon. This was Adonis, which made its famous close approach in 1936, after being discovered on February 12 of that same year by E. J. Delporte. Adonis was followed for two months, but was then hopelessly lost. However, in February 1977, 41 years later, a number of photographs were taken in the hope of re-locating it, and the programme was a success. The image of Adonis was found on one of his plates by Charles Kowal.

Even Adonis does not hold the record for the closest approach. This distinction goes to Hermes, a true Earth-grazer, which in 1937 winged past the Earth at a distance of only 780 thousand kilometres—only just over twice that of the Moon. The diameter of Hermes has since been estimated at no more than 1,000 metres, and so it is very tiny. It has of course been lost without a trace and will possibly never be found again, unless by pure chance.

Some close approaches by asteroids are expected in the not too distant future, but they will unfortunately all be pretty faint. Icarus is reckoned to approach to within 23 million kilometres of Earth in 1987, and the asteroid Toro to within the same distance in 1980. By far the closest approach within the next few years is expected to be made by Apollo. It will pass the Earth at a distance of only 9 million kilometres in both 1980 and 1982.

Possibly the most interesting asteroid, so far as the amateur is concerned, is Eros, or asteroid No. 433. Although when at aphelion it is slightly further away from the Sun than Mars, at perihelion it can approach to within 22.5 million kilometres of the Earth. This is not particularly close, compared to some, but its diameter is 23 kilometres and, though not large compared with the first four asteroids discovered, it is the largest of the close-approach asteroids and therefore appears the brightest. At its closest its magnitude may approach +7 and it can then easily be seen with binoculars. Such occasions for Eros are rare, but we were fortunate enough to experience one in 1975. Its January 13 opposition date was then only 11 days before its perihelion date and its closest approach occurred on January 23. Eros' last really favourable approach occurred in 1894, but it went unnoticed since the asteroid was not discovered until 1898. The 1975 approach was therefore the closest ever observed and was studied intently by amateurs and professionals alike.

In the past Eros has been used to obtain an improved value for the solar parallax, which is the angle subtended by the Earth's equatorial radius at a distance of one astronomical unit. The parallax is important for establishing the distances of objects within the Solar System. The more accurately its value can be determined the better will be the measurements of these distances. Eros' reasonably close approaches of the past have helped improve the accuracy, which makes it of historical interest from this point of view alone.

The 1975 approach carried an added bonus. When Eros was at its closest, on January 23, it occulted a fairly bright naked-eye star, Kappa Geminorum. The magnitude of this star is +3.6, and anyone lucky enough to be in the path of Eros' shadow would have seen the star blink out for about two and a half seconds. The width of the path would, of course, be the same size as the asteroids.

Observations of events such as these can often lead to quite accurate de-

Vesta among the stars of Leo, taken with a 135mm telephoto.

terminations of asteroid sizes. Unfortunately, if an observer were as little as 12 kilometres from the centre line of the shadow's path as it raced across the Earth, he would have missed the occultation and suffered the frustration of seeing the asteroid pass very close to the star without covering it. A number of observers in the USA, notably those in parts of Massachusetts and Connecticut, did manage to witness the splendid spectacle, but observers in Britain saw only a near miss, with the asteroid passing slightly to the west of the star. Nevertheless, as far as the whole apparition of Eros was concerned, Britain was a favourable vantage point. During its journey through the sky, which took only a couple of months, the asteroid passed through the constellations of Gemini and Canis Minor, making spectacularly close approaches to the bright stars Pollux and Procyon.

Fluctuations

Eros is also remarkable in that it displays rather rapid fluctuations in brightness. From these variations it has been established that the body rotates on its axis once every 5 h 16 m and that it is somewhat elongated in shape and rotates on the shorter axis. (It is sometimes described as a cylinder with rounded ends.) During the rotation period, there are two brightness maxima and two minima, making it possible to detect changes in brightness over a period of only 15 minutes.

The amplitude of its brightness variations may often be as much as $1\frac{1}{2}$–2 magnitudes. It is easy to estimate the brightness of this object by comparing it with certain stars in the same telescopic field. Find a star of similar magnitude and you will see that after a very short time Eros will grow either fainter or brighter than the star. If regular estimates are made, each 15 minutes over a 5 hr 15 min period, and the resulting values plotted on a graph against magnitude and time, an interesting light curve will emerge.

Observers using moderate-sized telescopes will not be able to resolve Eros into anything more than a star-like point. During one approach, in 1931, however, astronomers in South Africa using a 650-mm refractor reported that they were able to see the shape of Eros changing in time with its rotation period. For this type of observation a large first-class telescope and superb conditions of seeing are essential.

Many asteroids show brightness fluctuations, indicating that they are rather irregular in shape, but they are usually not as spectacular as Eros. There is one, however, that alters in

Photograph of Eros among the stars of Gemini. Its tiny image can be seen close to the bright naked eye star Pollux.

brightness by as much as 6 magnitudes, which is the largest amplitude of any asteroid. It is called Geographos, because it was found during the Palomar Sky Survey which was sponsored by the National Geographic Society. Though only small, it is very elongated, measuring 4 kilometres by 0.8 kilometres. This asteroid is expected to pass the Earth at a distance of only 13.5 million kilometres in 1983. It will, unfortunately, be rather faint.

Occultations of stars by minor planets, of which the Eros Kappa Geminorum event was a stunning example, are of extreme value. It is unfortunate that they are so rare. If the star and planet are bright enough, the amateur can play a useful role in the observation of these events. The more observers there are spread over the expected path of the occultation, the better the chance of obtaining some sort of result. The exact path is never known with certainty, so observers need to be spread over a wide area. We know the diameter and shape of an asteroid can be determined quite accurately by this method of observation, but there is always the chance of something unexpected, as happened on one occasion in 1978.

Double asteroid

It had been predicted that the fairly bright, 9.5 magnitude Herculina, asteroid 532, would occult a 6th magnitude star in Virgo on June 7 of that year. Herculina is quite a large asteroid, having a diameter comparable with that of Juno. The occultation was observed from a strip crossing Arizona, Nevada and California, and was seen quite well by observers using reflectors with apertures in the order of 100 mm to 250 mm. The occultations lasted on average 20 seconds, indicating a diameter of 243 kilometres, which is substantially larger than previous values obtained by other methods. This discovery alone made the observation worthwhile, but something else that was very odd happened. Just 90 seconds before the predicted occultation, sensitive photoelectric equipment recorded a brief disappearance of the star, lasting for just over 5 seconds. Similar results were obtained at two separate stations quite independently of each other, and so the secondary occultation seems to have been real. The diameter of this other, possible, object has been estimated at around 46 kilometres and its distance from Herculina, at that time, around 1,000 kilometres.

It is tempting to conclude that Herculina must be a double asteroid, or an asteroid with a natural satellite. There could, of course, be other explanations. There might, for example,

have been another asteroid in the same position—goodness knows there are enough of them. This would still be an incredible coincidence, and the possibility no more likely than the existence of a double asteroid. Alternatively, temporary light variations, in the star itself, could bring about the effect. The sky is full of 'variable stars', which continually alter their brightness. This star, however, had not been reported as variable before, and there is no other evidence to suggest that it might be. All in all, it was a very strange event and serves as a good illustration of just what can happen.

It is a fairly simple matter to record the images of minor planets by photography, which is quite possibly the best way for both the amateur and professional alike. Bright objects, such as Vesta, can be recorded easily with a standard 50-mm lens, working at about f2.8. If a fast film is used, say 400–500 ASA, it should be possible to record it with only a 15-second exposure. With such a relatively short exposure no guiding is needed, since the motion of the stars across the sky, due to the Earth's rotation, would not be sufficient to show in this length of time. For the fainter asteroids a longer exposure will be required and the camera must be guided. If you have a telescope with a drive, there is no problem. Just fasten the camera on to the telescope tube, piggy-back style, in a position that will not upset the balance of the instrument, and expose for as long as you consider necessary.

Hand-guiding

Even if your telescope has not got an electric- or clock-drive, guiding by hand should be possible, provided that the telescope is equatorially mounted. But you will need patience and a steady hand. Just fix cross-wires into a wide-angle, low-power eye-piece and look through the telescope with the eye-piece. Keep either the asteroid or, if this is not bright enough, a nearby star, centred on the cross-wire while the exposure is being made. You will find that it helps to put the focus of the telescope slightly out. This gives the guide object a small disc instead of just a point of light, so that it is not completely hidden by the wires. The author has hand-guided in this way for up to 45 minutes, which, though it may be tedious, is not difficult. Such a long exposure, however, is not usually necessary. With a fast film and wide-aperture lens it is possible to get down to magnitude 10 in only a few minutes. Good pictures can also be obtained in this way by using a telephoto lens of 135-mm or 200-mm focal length. A picture taken with this type of equipment during the 1975 Eros close approach is shown here. Even when Eros had dropped to 9th magnitude it still showed up well with exposures of less than 5 minutes.

Black and white photographs are fine, but you might also like to try colour. Transparencies are often very effective. You may, incidentally, if the asteroid is bright enough, and moving fast enough, try cutting down the aperture of the photographic lens and exposing for much longer, say up to one hour. You will then find that, instead of being a small point on the photograph, the asteroid will show as a short line. This will be the trail left by the asteroid as its position changed during the course of the exposure.

Before we leave the subject of minor planets there is one recent discovery that deserves mentioning, although, strictly speaking, it does not perhaps belong here. On November 1, 1977, Charles Kowal was examining some photographic plates, taken on the previous October 18 and 19 with the 122-centimetre Palomar Schmit Telescope. He noticed the trace of a faint moving object, whose magnitude was +18. This object, Chiron, as it is now called, was remarkable in that, although to all appearances an asteroid, it was found to be orbiting between Saturn and Uranus and not between Mars and Jupiter.

At perihelion Chiron is 1,300 kilometres from the Sun, or just inside the orbit of Saturn. At aphelion, however, it is a staggering 2,800 million kilometres from the Sun, almost as far away as Uranus. It cannot therefore be classed as one of the minor planets discussed in this chapter. Its diameter is thought to be approximately 450 kilometres, though this is at present uncertain. In size it resembles the larger asteroids, and it has been suggested that it might be the larger member of a trans-Saturnian asteroid belt. Only time will tell.

Asteroids are fascinating little worlds and as yet no spacecraft has paid any of them a visit, although it has been suggested. We can imagine what they look like from close-range since it is possible that the two tiny moons of Mars are nothing more than captured asteroids. If not, they must surely resemble them. The views from certain asteroids would at times be quite spectacular, especially from those passing close to the major planets.

Perhaps asteroids are themselves the remains of a planet that for some reason met a violent end. Some theorists try to support this idea by citing the fact that certain asteroids are rapidly turning end over end. Could they not even be left-overs from the formation of the Solar System? Whatever may be the true explanation, there is plenty of interest that can be found in what many astronomers consider to be the litter of the Solar System.

Icarus and the Sun. Sometimes this asteroid approaches the Sun so closely that it may glow red hot.

Jupiter

As we leave the asteroid belt and move further away from the Sun we will eventually come upon Jupiter. This planet is totally different from any we have so far encountered. The four planets we have already described, although different in detail, are basically similar to each other in their make-up. All of them have a comparatively large metallic core surrounded by a rocky mantle and finally, with the exception of Mercury, a thin rocky crust.

They are also of a similar order of size and form a tightly knit group, quite close to the Sun. For this reason they are classed as the terrestrial planets, in recognition of the fact that they resemble the Earth in size and density. The next group of planets, again numbering four, while resembling each other, in no way resemble the terrestrial planets. These are the giant gas planets of which Jupiter is closest to the Sun and the largest member of the group which are referred to as the 'Jovian planets'.

The largest planet

Jupiter is the largest of all the planets. Its equatorial diameter is 142,200 kilometres, which is more than 11 times the diameter of Earth. Its volume is 1,310 times the Earth's, which means that this number of Earth-sized bodies could be crammed inside a body the size of Jupiter. In fact, Jupiter's volume is greater than that of all the other planets put together. Despite this the planet's mass is only 317.9 times the Earth's mass, so 318 Earths would weigh slightly heavier than Jupiter. This tells us that the density of the planet is very low, which is a distinguishing feature of the gas giants. The actual density of Jupiter is only 1.3 times that of water, whereas the Earth has a density 5.5 times that of water. All the terrestrial planets have densities ranging from 4 to 6 times that of water. The gas giants have densities which are, on average, only 1–2 times that of water.

Despite its size, Jupiter has the most rapid spin of all the planets. Its axial rotation period, at the equator, is only 9 h 50 m 30 s. Because of this rapid spin and low density, the planet's figure is distorted—it literally bulges at the equator and it is flattened at the poles. Jupiter's polar diameter is 134,700 kilometres or 7,500 kilometres less than its equatorial diameter. The tilt of its axis is only 3° 04′ to the plane of its orbit. Since this axis is almost erect, this means that Jupiter has no seasons like Earth and Mars. The orbital period of Jupiter is 11.86 years and this means that there are something like 10,530 of its short days in its year.

The mean distance of Jupiter from the Sun is 778.34 million kilometres and, although the planet has a rather small orbital eccentricity, its actual distance from the Sun can vary by as much as 74.8 million kilometres. The closest that Jupiter can ever approach to the Earth is 588 million kilometres and this must be at a perihelionic opposition. At its most distant conjunctions, however, its distance increases to 967 million kilometres. Yet although Jupiter is so far away it always presents a sizeable disc for telescopic study. The mean angular opposition diameter, measured across its equator, is almost 47″, although when it is at its closest, this value may exceed 50″. Even at its smallest it is never under 30″ and a minimal amount of magnification will resolve it. A good pair of 10 × 50 binoculars are usually sufficient.

A fairly small telescope will give very good results. Even a 75-mm refractor will show the slightly flattened, golden-coloured disc, marked by dusky belts parallel to its equator. These belts are the major feature of the disc. A 150-mm reflector will show some considerable detail within these belts and in the lighter zones between them. In fact, very useful work can be carried out with an instrument of this size. However, as with all the planets, the bigger the telescope the better. A 300-mm reflector will show a wealth of detail. In fact, at times there is too much to hope to be able to record.

Atmospheric features

The dusky belts and light zones which mark the planet are features of the atmosphere. We cannot see the surface—if, indeed, it has one. Telescopes of moderate aperture reveal a considerable amount of complex detail. Dark and light spots, wisps, streaks and dusky filaments can always be seen and the study of their motions tells us much about the currents that exist in the upper atmosphere of Jupiter.

One of the main objects of any programme aimed at observing Jupiter is the determination of longitude of as many observed features as possible. This gives us an insight into the various drift rates of features on certain parts of the planet. From this it has been found that most of the belts and zones have their own individual rotation periods. There are, however, basically two main systems of rotation in the planet's atmosphere, the equatorial regions having a rotation period of 9 h 50 m 30 s while the tropical, temperate, and polar regions rotate in 9 h 55 m 41 s. These systems are referred to as I and II respectively.

Nomenclature

The method used to decide the longitude of a given feature is the same as that described for Mars. An imaginary line through the centre of the disc, passing through the poles, marks the central meridian. When a feature is estimated to be exactly on this line a note of the time should be made, as accurately as possible. For large features the preceding and following ends, as well as the centre, should be timed across the CM, or central meridian. With Jupiter the speed at which the features are taken across the disc is quite rapid and motion is easily perceptible over a period of five minutes. With practice it is possible to assess, to within the nearest half minute, when a feature is actually on the CM. Tables of longitude for the CM, at 0 h Universal Time daily, will be found in various astronomical handbooks, together with changes of longitude in intervals of time, down to one minute, as we have described for Mars. There are, however, in this case two tables, one for each system of rotation.

By continual study of the various features in this way many currents are revealed within each system and occasionally features will be found that have either very rapid or very slow velocities compared with their surroundings.

The nomenclature of the planet's

Imaginary view of Jupiter from Amalthea—a spectacular sight which would fill one-quarter of Amalthea's sky.

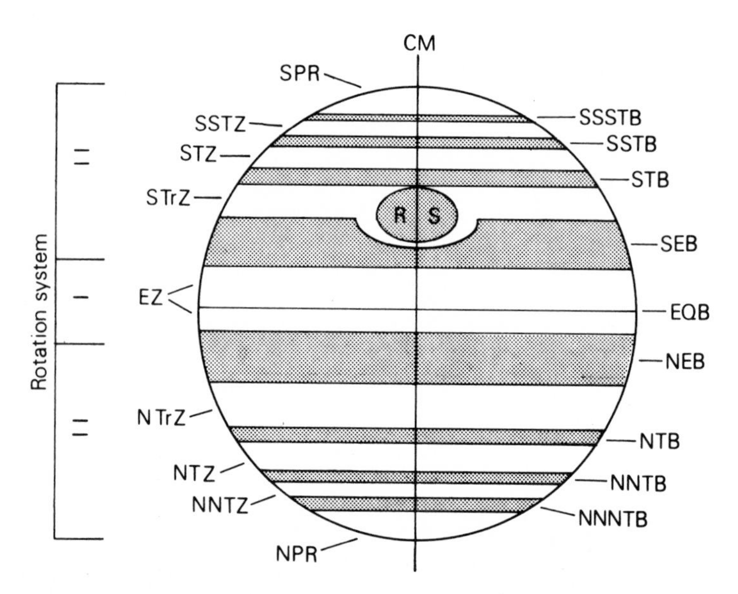

Jupiter's belts and zones and the two rotation systems.

belts and zones is fairly simple. The most consistent features are the Equatorial Zone (EZ), which usually appears as a light band around the centre of the planet. At the centre of this will usually be seen a dusky line called the Equatorial Band (EB). This main EZ is flanked to its north and south by prominent dark belts known as the North Equatorial Belt (NEB) and South Equatorial Belt (SEB). Moving towards the poles, we will then encounter bright North and South Tropical Zones (NTrZ and STrZ). Still further on we then come to the north and south temperate regions. These regions comprise a number of less prominent, and somewhat variable, belts and zones. First and most prominent are the actual North Temperate Belt (NTB) and the South Temperate Belt (STB), which are in turn bordered by the fairly bright North Temperate Zone and South Temperate Zone (NTZ and STZ).

As we begin to draw closer to the actual polar regions things begin to get a little confused, belts and zones becoming rather poorly defined. If features are seen, these progressively become NNTB, NNTZ, NNNTB, NNNTZ and so on. (For southern features S is substituted for N.) Finally we come to the actual North Polar Region and South Polar Region (NPR and SPR). This completes the basic nomenclature. Some belts, however, may be seen to be split into two parts, their north and south edges being separated from each other by a narrow light region. In this case the belt in question will be described as having a north and south component which, for the NEB, say, would be written NEBn and NEBs.

Belts and zones

As for the rotation systems, System I includes all parts of the planet between the n edge of the SEB and the s edge of the NEB. This region, of course, incorporates the EZ and is generally referred to as the 'Great Equatorial Current'. System II incorporates all regions s of SEBn and n of the NEBs, thus including all tropical, temperate and polar regions. Included below is a brief description of each region, together with its general appearance.

NPR: Usually slate grey and featureless. In the absence of any temperate belts this may appear quite extensive, reaching the latitude of the NNTB. On average, it extends to a latitude of about $+48°$. Occasionally very delicate zones or belts may be detected, leading to some variability in the intensity of the region. Spots have been detected but are rare and always difficult. The area appears featureless, mainly through foreshortening and atmospheric absorption. The average rotation period for the area is 9 h 55 m 42 s.

NNTB: This is very variable in appearance and always difficult to observe. It often appears broken and discontinuous. The average rotation is 9 h 55 m 20 s.

NNTZ: This usually has a similar intensity to the NPR.

NNTB: More often than not this belt will appear in separate sections around the planet, as shown on some of the drawings presented here. It can often appear quite wide. The average rotation period is 9 h 55 m 42 s.

NTZ: Greatly variable in width and brightness, sometimes appearing as one of the brightest zones. On occasions quite prominent light spots will be detected.

NTB: A very variable belt indeed, sometimes faint and almost invisible, sometimes very prominent. It also varies in width, as can be seen from some of the drawings here. Sometimes the belt might be seen to have two or even three components, also many small dark spots or streaks may be seen. The strange thing is that, although the belt is part of System II, it often displays features which have rotation periods similar to System I.

The colour of the belt is also worth noting. Often it appears grey or brownish, although sometimes it may have a strong reddish colour and be very striking in appearance.

NTrZ: This is usually a brilliant zone, greatly varying in width. It is also the scene of a fair amount of activity. This is not really surprising since it is situated between two very active regions. To its north is the often disturbed NTB described above, but to its south lies the violently active NEB.

Disturbances here are associated with the North Temperate Belt's occasionally violent outbreaks. The average rotation period for this area is 9 h 55 m 29 s.

NEB: This is a prominent, and often the most prominent belt. It is nearly always active, particularly along its southern edge. Usually it consists of a north and south component and it can be quite variable in width. There have been occasions when the belt has faded considerably, but this is a rare phenomenon. Dark spots and streaks are usually seen along its northern edge and sometimes small brilliant spots seem to eat into this part of the belt. The general region rotates in 9 h 55 m 29 s and, along with the NTrZ, this is classed as the North Tropical Current.

The southern edge of the belt is an intensely active region. Large dark and bright plumes issue into the EZ, and bright spots at the roots of the light plumes cause large dents in the southern edge of the southern component. These features are very prominent and are easily seen with small telescopes. The rotation of this region is part of System I and is thus classed as part of the Great Equatorial Current.

21.XI.75 20.21 UT 254mm Refl ×300 The markings of Jupiter comprise prominent and detailed cloud belts and bright zones. The Great Red Spot is seen here at one of its prominent periods.

Southern regions

The central region of the belt can often be quite light and the rotation periods of any spots seen in the region are difficult to establish. Any break through the whole belt soon becomes drawn out since the southern portion, having a more rapid rotation period, quickly leaves the northern portion behind.

EZ: This is usually a broad bright zone, although in recent years the southern half has been quite dark—in fact, at times it has been as dark as the equatorial belts. Large spots and patches, both light and dark, mark the region and these are occasionally very complex in shape. The northern half of the zone usually exhibits the most spectacular features.

At the centre of the EZ is a narrow belt called the 'Equatorial Band'. This is very irregular in intensity and is often greatly disturbed by the turbulence of its surrounding regions.

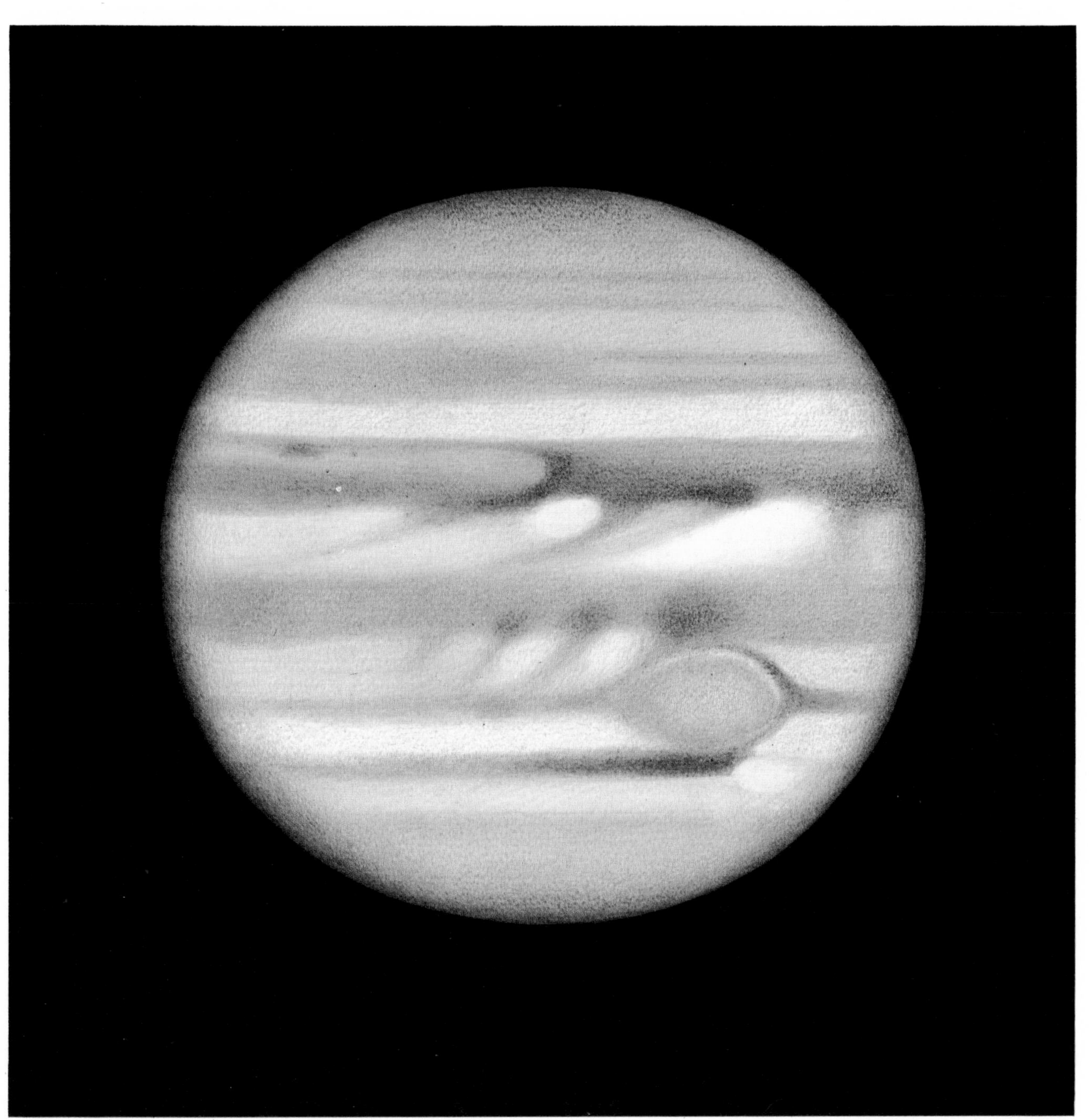

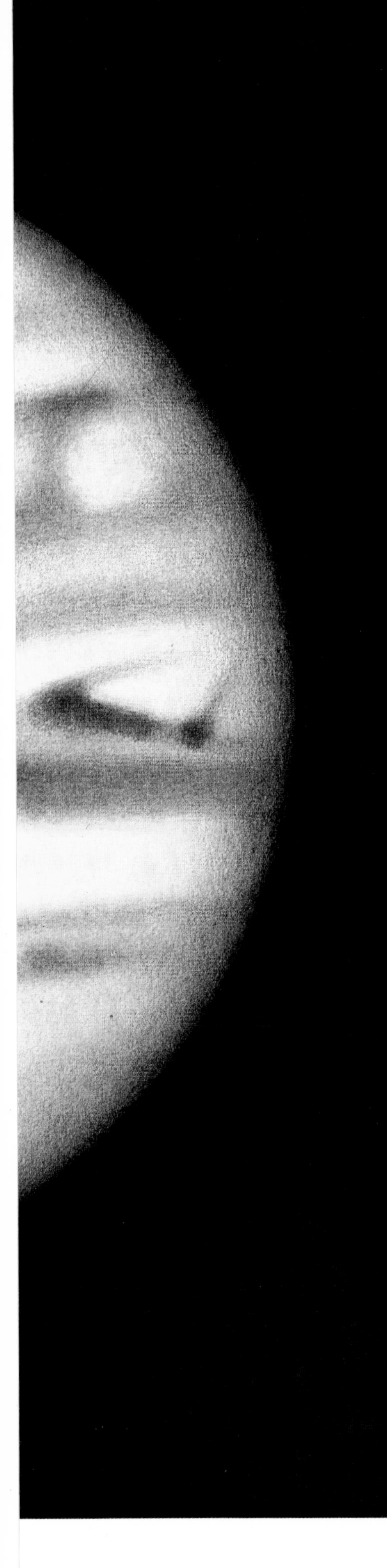

SEB: In this belt we have one of the most violently active regions of the planet. Indeed, the periodic development of this belt is so complex and so interesting that it is necessary to discuss it in some detail. We do this in the part of this chapter concerned with features of particular interest. The belt is a counterpart to the NEB, but is nothing like as consistent in intensity and prominence. Its northern edge forms part of the Great Equatorial Current. Usually the southern part of the belt has a rotation period similar to that for System II, being 9 h 55 m 39 s. However, during active periods, things conform less well.

STrZ: This can often be the brightest zone on the planet, and its colour is usually white or creamy white. In this zone is situated the famous Great Red Spot, which must be classed as the strangest feature on the whole planet. It has also been the site of a rather interesting disturbance known as the 'South Tropical Disturbance'. Again, both this and the Great Red Spot (GRS) will be discussed in detail later in the chapter. General rotation period for the region is 9 h 55 m 36 s.

STB: This is often a fairly prominent narrowish belt, sometimes appearing double. It has been reported as second in prominence to the NEB from time to time. Very dark and quite large spots may be seen on the belt and occasionally whole sections of the belt seem to disappear while other sections remain prominent. Along the belt's southern edge, three quite large white oval-shaped features are at present situated. Their northern parts trespass some way into the STBs while the southern parts of these ovals encroach into the STZ. Dark knots mark their preceding and following ends. The general rotation period of the belt is 9 h 55 m 20 s.

STZ: This is quite a wide zone at times. White spots are a less common occurrence. Sometimes a faint grey line may be noticed, appearing as an extra component to the STB. Sometimes, however, it seems to cross the zone diagonally. The average rotation period for the region is 9 h 55 m 20 s and, together with the STB, the region forms the South Temperate Current.

28.XI.76 23.58 UT 419mm Refl ×300 At times when the South Equatorial Belt of Jupiter undergoes one of its spectacular revivals the Great Red Spot may lose prominence and colour.

SSTB: This is usually faint, forming an ill-defined border to the south polar shading.

SSTZ: This is often a dusky and difficult zone. Both the SSTB and SSTZ form the South South Temperate Current and have an average rotation period of 9 h 55 m 7 s.

SSSTB: This is rarely seen as, in fact, is the SSSTZ. When these features are recognized they indicate a current similar to the SPR.

SPR: This is a slate-grey and featureless area. The rotation period has been virtually impossible to establish by Earth-based observation but it seems to be in the region of 9 h 55 m 30 s.

Every shade of colour

The colour of the planet is worth special mention since almost every shade has been reported at one time or another. The general appearance of the planet can be quite deceptive on first inspection, appearing creamy white with greyish belts. There are, in fact, many who never see a great deal of colour on the planet, but those who are fortunate enough to catch fleeting glimpses in near perfect seeing are taken aback by the amount of, sometimes vivid, colour present. True, most of the zones do range from white to yellowish white, but they can sometimes appear strong yellow, chrome yellow or even biscuit. With close inspection, it will be noticed that the zones vary quite appreciably in their tone and texture. The belts on the whole appear grey but, again, close inspection will reveal many shades of grey, brown and reddish brown. Sometimes narrow streaks running the whole length of a component will have a very strong red, or even fiery red, colour. Sometimes the NEBn will exhibit a clear vermilion streak around the whole circumference of the planet. Dark spots seen on this belt may appear very deep maroon, while other spots or streaks in the belt may be bluish grey or even indigo.

Reports often indicate the NEB as being of a copper colour. The NTB seems as variable in its colour as in its intensity. Sometimes it appears red-

ence in motion is brought about by the fact that the SEBn is in System I, while the SEBs is in System II but, for the purpose of describing the motions of these spots relative to each other, System II is used for both regions.

When the retrograding SEBs spots eventually reach the preceding end of the GRS a fade of the latter will usually begin. During the disturbance the bright SEBZ becomes filled with a wealth of detail, in the form of dark and bright spots, wisps, festoons and chunks of dark material. Eventually the region becomes a conspicuous dark belt, often rivalling, or even surpassing, the intensity of the NEB. Then the SEBs will begin to drift slowly north, resulting in a narrowing of this belt. Finally, a slow fading will begin. The region eventually returns to its former brilliance, preparing the way for a fresh revival.

This, in essence, is what may usually be expected of a typical SEB revival, if any can be regarded as typical. There have been a number of such outbreaks since the first recorded one of 1919. The second one observed occurred in 1928 and this was possibly the most spectacular. Further disturbances were recorded in 1943, 1949, 1952, 1955, 1958 and 1962. It seemed from this that some sort of periodicity was setting in since, following the 1949 eruption, a fresh disturbance had occurred after a period of three to four years. In true Jupiter fashion, this pattern was disrupted with a long gap from 1962 until 1971, when there was a further great upheaval. The last one, observed in 1975 was almost as violent as that of 1928 and it will be noticed that this again fitted in to the three-to-four-year periodicity cycle.

Individual characteristics

Although these outbreaks follow a general pattern, each has had its own individual characteristics. Sometimes the revival has had two sources of outbreak instead of one. The great revival of 1975 had an unprecedented four separate sources. Other things

Above left *2.IX.75 00.30 UT 254mm Refl ×300 Rapidly moving spots on the darkening SEBs.* Left *3.IX.75 23.32 UT 254mm Refl ×300 Detail in the belt's central region.*

which set this particular revival apart were the virtual absence of direct moving spots on the SEBn and the fact that the GRS only lost prominence on its northern part. A series of drawings presented here show various stages of the 1975 revival.

Great upheaval

An interesting hypothesis, by American astronomer E. J. Reese, suggests that outbreak sources of the SEB revivals may be connected to permanent features of Jupiter's actual solid body below the cloud deck. A system of rotation (System III derived from radio observation) probably relates to the rotation of this solid body and it fits in with the drift rate of the initial source features. It has been found to fit in with previous disturbances and it happened that the 1975 sources were very close to their predicted positions, based on this third rotation system.

Further observations are vital if this is to be cleared up, but information derived so far is yet another brilliant illustration of what visual work can achieve.

Apart from the regular and periodic changes, sometimes great and quite unexpected upheavals occur. In 1959, the colour of the EZ underwent a dramatic change. Until then its colour had usually been bright creamy white. Suddenly it became dark yellow, with some observers reporting an orange, or even pink, tone. In the years that followed, the zone gradually darkened as an apparently new dark belt formed in the region. This, together with the Equatorial belts, gave the appearance of a very wide, dark, central band, in which there were a great number of dark and bright streaks and spots. During 1966 and 1967, there was a temporary return to normal, with the zone again becoming bright. However, a further darkening set in during 1968. The

Above right *30.X.75 21.12 UT 254mm Refl ×350 A build-up of white areas around the Red Spot which is then likely to fade. Note darkness of SEB following the Red Spot.* Right *31.X.75 20.40 UT 254mm Refl ×350 SEB detail shows revival is under way. Note the bright white oval feature of the STB.*

Above *9.XII.76 19.01 UT 419mm* Right *9.XII.76 19.51 UT Both 419mm Refl ×300 Jupiter's rapid rate of spin. The second picture shows how markings are displaced to the left.*

zone retained a strong dark reddish brown colour until 1973, when another major change came about.

In that year, huge brilliant white plumes appeared on the northern half of the zone, their sources appearing as even brighter small white spots, situated on the southern edge of the NEB. Two of the plumes were particularly prominent. These were found to be moving with great speed compared with other features in the region. This resulted in their catching up with other features and, in so doing, they appeared to devour everything with which they came into contact. Other smaller plumes were obliterated as the two large ones passed, but fresh plumes seemed to develop in their wake, forming very quickly in the confusion following the great plumes.

In 1975 both of these features disappeared in less than a week, leaving behind them a scene of tremendous upheaval. New plumes formed during 1976 and these have since been a major feature of the equatorial regions. The zone has generally continued to brighten up and this may possibly result in Jupiter returning to its pre-1959 appearance.

Three white ovals

Situated on the southern edge of the STB, and encroaching into the STZ, are three rather large white oval areas. Although these are in no way violently active and spectacular, like the disturbances we have already mentioned, they do have some points of considerable interest.

The spots themselves were first noticed as such at the end of the 1940s. Before then the belt seemed to have three short streaks, each separated from the other by some distance.

These streaks received the designations AB, CD, EF. During the 1940s the streaks lengthened and the gaps between them became shorter. By 1950, the region of these short gaps brightened up to result in the appearance of the white ovals. This is reflected in the present designations assigned to the ovals FA, BC, DE. These letters still, in effect, refer to the preceding and following ends of dark sections of the belt rather than to the bright features.

The ovals have remained a prominent and familiar feature since 1950. An interesting thing about them is that although they display individual rates of drift that differ from each other, there has never been an occasion when two of the ovals have come into contact, which is what one would expect. As soon as they come into close proximity to each other there is an effect on their motion which suggests some sort of repulsion force operating between them. Just what this is we do not fully understand. Very interesting conjunctions between these features and the GRS are observed since these ovals move at a faster rate than the latter.

The atmosphere of Jupiter can present many incredible phenomena. Possibly one of the most puzzling was observed during the period of the existence of the South Tropical Disturbance. In 1920, it was noticed that spots moving in a retrograde fashion, slightly south of the SEBs (such as those described as occurring during the SEB revivals), seemed to disappear on meeting the preceding end of the South Tropical Disturbance.

Circulating current

However, a few days later similar spots would be seen near the northern edge of the STB moving in a direct fashion. Observations of a similar type through subsequent years implied that there was a circulating current operating in the STrZ. The disappearance of the South Tropical Disturbance around 1940 resulted in no further observations of this type being made. There is, of course, always the chance of a recurrence.

Certain spots have been observed to change the direction of their drift quite considerably. Such spots, again situated on the southern part of the STrZ, appeared to have a fairly rapid direct motion when first observed. This motion would, however, slow down until the spots not only became stationary, in relation to their immediate surroundings, but displayed a slight retrograde motion, only to begin a direct motion again after a very short time. Such were the famous Oscillating Spots seen in the 1940s.

Below *6.XI.76 22.05 UT* Right *13.II.77 19.40 UT Both 419mm Refl ×300 The gradual change in appearance of a region in the planet's NEB and EZ. A lapse of three months shows how long-lived certain features can be, even looking at the atmosphere and not features of the actual surface.*

Atmosphere in turmoil

The visible atmosphere is evidently in a constant state of turmoil and there is enough happening to keep any observer happy. Jupiter has another added attraction—its thirteen satellites. Very many of these satellites are too faint to be detected with amateur telescopes, but four are very bright and conspicuous objects. These four 'Galilean' satellites, so called because it is generally recognized that Galileo was the first to draw attention to them, are easily seen with a good pair of binoculars. There are even numerous reports of their having been detected with the unaided eye. Since their opposition magnitudes are between +5.5 and +4.5, theoretically, they are well above limiting naked-eye magnitude. Were it not for their close proximity to the brilliant Jupiter they would be easy objects without any optical aid, but the brightness of Jupiter makes them a severe test of eyesight. Any keen-sighted person

wishing to try their detection in this way might find it a great help to blot out the image of the planet with a telephone wire or something similar.

When these bright satellites were first discovered in 1610 they gave the first visual proof that the Earth is not the centre of the Universe. Here were four objects revolving around Jupiter and not the Earth. They have also been used in connection with the determination of the speed of light.

Jupiter's satellites

The satellites themselves are quite large. Satellite I (Io), the closest to Jupiter, has a diameter of 3,652 kilometres and is slightly larger than our Moon. Its actual distance from Jupiter is 422,000 kilometres, giving a maximum angular opposition distance from the centre of Jupiter of 59.4″. Its stellar magnitude is +4.8. Satellite II (Europa) is a little smaller than our Moon, having a diameter of 2,900 kilometres. Its distance from the centre of Jupiter is 671,000 kilometres, yielding a maximum angular opposition distance of 2′ 18.4″; its magnitude is +5.2. Satellite III (Ganymede) is the largest of the four, with a diameter of more than 5,000 kilometres, making it larger, in fact, than the planet Mercury. Its distance from the centre of Jupiter is 1,073,000 kilometres, giving a maximum angular opposition distance of 3′ 40.1″. Its magnitude is +4.5, making it the brightest. Satellite IV (Callisto) is a large body, almost the size of Mercury, with a diameter of 5,000 kilometres. Its distance from Jupiter is 1,884,000 kilometres. This gives it a maximum angular distance of 10′ 17.6″, or about $\frac{1}{3}$ the angular diameter of our Moon. Its stellar magnitude is +5.5, making it the faintest.

As for visual observation of the satellites, they are quite easily resolved into discs with only moderate-sized telescopes, since their angular diameters at opposition are quite appreciable. Io is just under 1″. Europa is less than 0.5″, and is there-

fore the most difficult to resolve. Ganymede and Callisto both have angular diameters in the region of 1.5″.

To detect any markings on these satellites requires quite large telescopes and exceptional conditions. There is a chance that something may be seen with telescopes in excess of 450 mm and the best time to try to spot surface features is when the satellite is entering, or leaving, transit across Jupiter. It will then be seen as a sharp disc projected on the dark limb of the planet, with the glare from its tiny, but bright, image much reduced. Markings that have been detected are in the nature of vague dusky patches and small bright areas.

Satellite phenomena

Io has a yellowish orange colour, though this is difficult to detect visually. It appears to have a bright central band with darker poles. Europa has a brilliant white colour. It has particularly brilliant poles and signs of a darker equatorial band. Markings on Ganymede have in the past suggested that it may have an atmosphere. Clear dark patches have been detected, particularly on its equatorial regions, but some brilliant patches have also been seen that at times appear to obscure some of the darker markings. Callisto is very dusky and markings are difficult to make out. The fact that the markings of these satellites seem to be repeated when they occupy the same parts of their orbit led to the belief that they have 'captured' rotations. In other words, they tend to keep the same face towards Jupiter all the time.

The most fascinating aspect of these satellites is the actual phenomena presented by them while passing behind, or in front of, Jupiter. Observations can be made of eclipses, as the satellite moves into the shadow of Jupiter, of occultations, as the satellite moves behind Jupiter itself, of transits, as the satellite moves in front of and across the face of Jupiter, and shadow transits, when the actual shadow of a satellite falls on to the disc of Jupiter.

Between conjunction and opposition a satellite will be seen to enter into eclipse before reaching the preceding limb of Jupiter, followed by its emergence from occultation at Jupiter's following limb. The point at which the satellite enters into Jupiter's shadow will be close to the preceding limb of the planet, shortly after conjunction, but this distance will gradually increase to reach maximum separation at quadrature. Then the separation will gradually decrease until opposition, when the shadow cast by Jupiter falls directly behind and is, in fact, hidden by the disc so that, to all intents and purposes, occultations and eclipses appear simultaneously at opposition. In the case of shadow transits between conjunction and opposition, the shadow will always precede the satellite, with the greatest separation between the two again occurring at quadrature. Between opposition and conjunction the whole situation is reversed, with occultations preceding eclipses and shadow transits following satellite transits. Transits themselves always start at the following limb.

16.II.77 18.10 UT 419mm Refl ×300 Jupiter's larger, Galilean satellites. Note how these align with the belts of the planet.

For events concerning Io, eclipses and occultations are always merged because of the satellite's closeness to the planet. For the more distant satellites, Ganymede and Callisto, occultations and eclipses only merge around opposition. When Jupiter is at first quadrature each of these two satellites will be seen to enter and leave eclipse before reaching the planet to be occulted. At second quadrature these satellites will be occulted and then sometime later will enter into eclipse. Europa is an extreme case. Usually, like Io, occultations and eclipses merge. There are rare occasions, however, with Jupiter at quadrature, or a few days either side, and with the tilt of Jupiter's axis at its maximum with respect to the Earth, when the two events will just about be separated. This happened during 1976. On January 29 in that year, Satellite II appeared from occultation and almost as soon as it became fully visible its brightness started to drop rapidly with emergence into eclipse. From the first sign of the satellite until the final point of light had vanished took no more than eight minutes. It was a wonderful spectacle.

For shadow transits occurring close to opposition the shadow will appear very close to the image of the satellite responsible for it. At certain transits occurring at the exact time of opposition a satellite may even partially occult its own shadow. This event occurs most frequently with Io, but it is nevertheless quite rare.

The actual length of time an event will last varies from one satellite to another. Io and Europa complete their events quite quickly. Their synodic periods are 1 d 18 h 28 m and

3 d 13 h 17 m respectively and so their motion around Jupiter is quite fast. Io will complete a full central transit in slightly over 2 hours while its actual ingress and egress times (the time that the satellite takes to enter on to, or leave, the disc of Jupiter) are just over $2\frac{1}{2}$ minutes. Respective times for Europa will be almost 3 hours for a full central transit and slightly over $2\frac{1}{2}$ minutes for ingress or egress. Although Europa moves more slowly than Io it is a smaller body and this results in the two satellites having similar times for ingress and egress.

Ganymede and Callisto

For the two more distant satellites, Ganymede and Callisto, events take place at a much more leisurely pace. The synodic times for these two satellites are 7 d 3 h 59 m and 16 d 18 h 5 m respectively. This means that a full central transit for Ganymede will last in the order of $3\frac{3}{4}$ hours while that for Callisto will last for almost 5 hours. Both satellites are quite large, resulting in ingress and egress times of about 7 minutes and 9 minutes respectively.

The above times are for events crossing the central part of Jupiter's disc. Times can vary considerably as Jupiter's axial tilt alters with respect to the Earth. Events for Ganymede can occur quite close to the planet's polar limb and can therefore be of much shorter duration. At the same time, though, ingress and egress times may be lengthened, depending on the angle at which the limb of the planet is bisected. The extreme case here is Callisto. For a few years, around the time that Jupiter presents its maximum tilt to us, the latter satellite will pass either above or below the planet and no phenomena will be observed. On the whole, then, events for this satellite are quite infrequent.

When satellites and their shadows are in transit they each have their individual characteristics. Io, for instance, will look like a faint grey spot if superimposed on a bright part of the planet, but when crossing a dark belt it sometimes appears as a small elongated bright spot. This is a result of the satellite's bright equatorial region. The shadow of this satellite is very sharp and black. Europa has the highest albedo of all the satellites and, when in transit over a bright zone, it will be difficult to detect. When over a dark belt, however, it will be seen as a small brilliant spot. The shadow of this satellite, being largely penumbral, is not so black and is very fuzzy in outline. It is the most difficult of all shadows to see but it is still an easy object for small to moderate telescopes to be able to resolve.

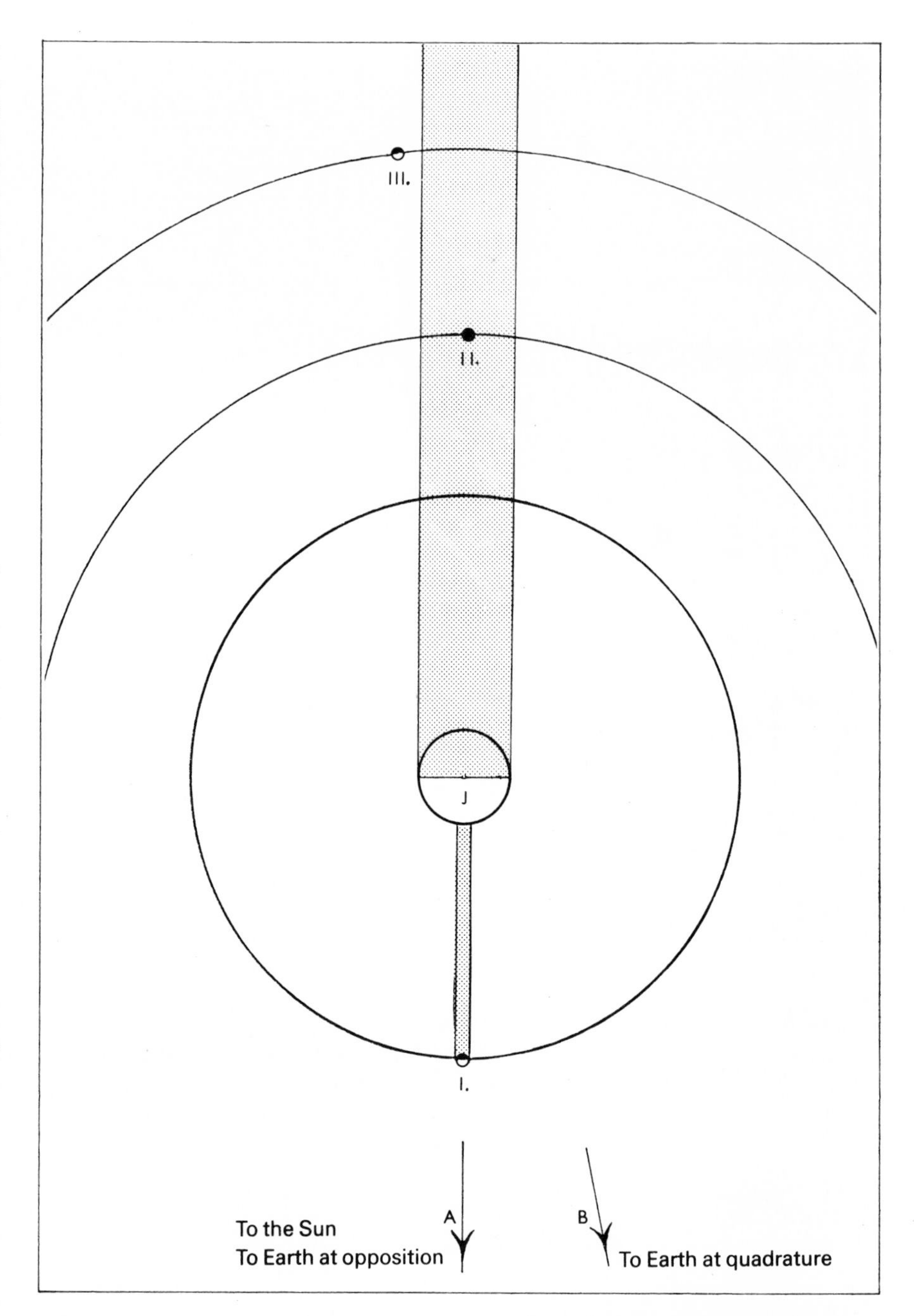

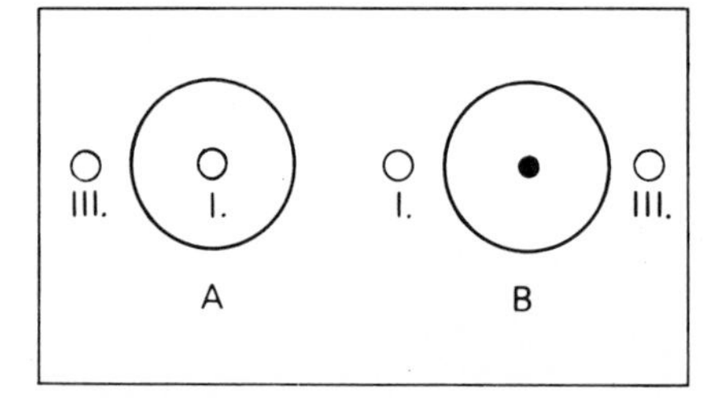

Above *A plan of the orbits of three of Jupiter's satellites, giving rise to certain phenomena. The shaded areas represent the shadow cast by Jupiter and by Satellite I.* Left *A diagram showing how these satellites appear from two different viewpoints, A and B. At A, Satellite I is occulting its own shadow; at B, Satellite I is well to the left of its own shadow and Satellite III can be seen between occultation and eclipse.*

Right *28.XII.76 18.44 UT 419mm Refl ×300 Occultation of Europa.*

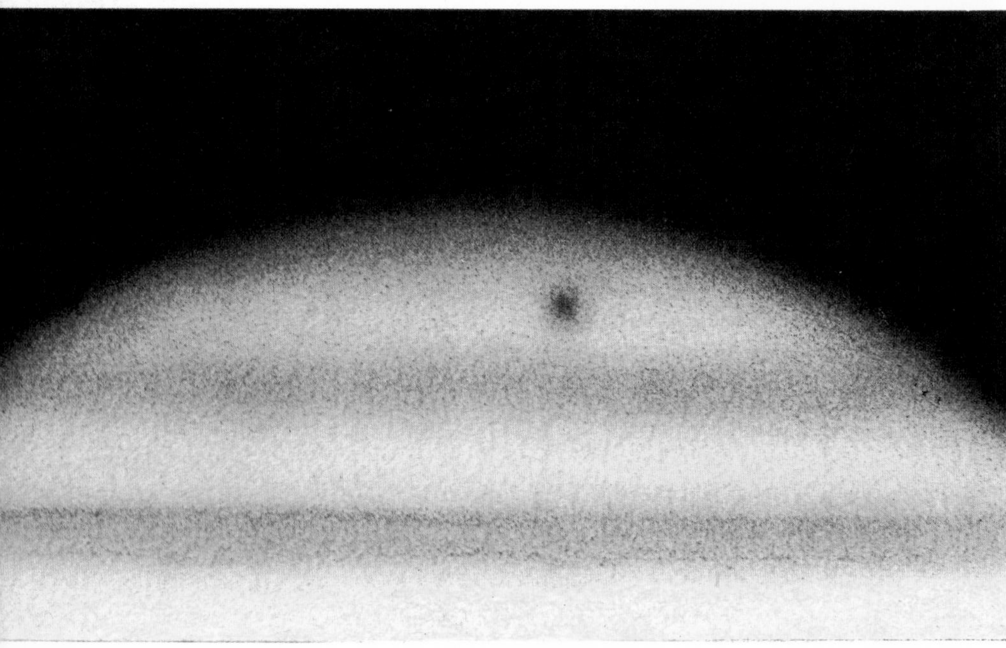

Ganymede is interesting when in transit since it appears very dusky and quite large. It is particularly interesting to watch an ingress of this satellite. As soon as it enters on to Jupiter's disc it will look like a bright round spot in contrast to the darkened limb of the planet. It is soon lost to view as it moves on to the disc and after twenty minutes or so will be noticed as a dusky spot, especially if seen over a bright zone. Detailed observations with large telescopes show this satellite as a wedge-shaped dusky streak, resulting from a dark equatorial feature on the satellite. Continued observations of this feature over many years have shown it to alter in appearance very slowly and the inference is that the rotation of Ganymede is not quite a captured one. The shadow of this satellite is very large and very black, and is the most conspicuous of all shadows and can be detected with only a 50-mm refractor. Transits of Callisto are the most beautiful of all, since it will be seen to turn from a bright spot to an intensely dark one as soon as ingress is complete. When in transit over a bright part of the planet, its silhouetted disc may be so black as to allow it to be mistaken for a shadow transit. This acts as a wonderful illustration of the varying albedos of these satellites and helps one to understand why, when Callisto is so large, its visual magnitude is below the very much smaller Europa. The shadow of Callisto is also black and sharply defined.

Mutual occultations

If the reader thinks all this is fascinating then consider the even greater fascination offered by the mutual phenomena of these satellites. At certain parts of Jupiter's orbit the Sun and the Earth will lie exactly in the plane of the satellite's orbits. Such times are spaced just under six years apart and, at best, last through one complete apparition. Then the satellites can either eclipse or occult each other. These events can be truly captivating.

The last two opportunities to ob-

27.XI.75 21.20–22.10 UT 254mm Refl ×300 Satellites I and III appear bright when they first move on to Jupiter's disc. I is later lost and III is a dusky spot.

serve them were afforded during the 1973 and 1979 apparitions. Mutual occultations and eclipses may be either total or partial, although the former are of course, quite rare. The duration of the events is again subject to great variation. Events concerning Io and Europa can be completed in five or six minutes. With eclipses, to witness the eclipsed satellite decline in brightness to almost, or even complete, invisibility and then rise again to its former brightness, all in the space of a few minutes, is a splendid sight. With large telescopes it may even be possible to see the partial phases clearly, with the shadow of one satellite taking a bite, as it were, out of the disc of another.

Triple shadow transits

Occultations also have interest, especially if the small bright disc of Europa passes over the dull disc of Callisto. Some events may be rapid but, of course, there are others which will be very drawn out. This depends on the relative motion of the two satellites involved. If they are moving in opposite directions then obviously events will proceed much more quickly than if the satellites are moving in the same direction.

There are times when the ordinary type of satellite phenomena may be more spectacular than usual, since more than one satellite will often be involved. It is not uncommon, for instance, to see two shadows on the disc of Jupiter at the same time. It may even be possible, although extremely rare, that three shadow transits will occur simultaneously. For this to be possible the shadow of Callisto must be one of those involved, the reason being that there is a simple relationship between the mean daily motions of Io, Europa and Ganymede resulting from the perturbations these three satellites exert on each other. This means that the inner three Galilean satellites cannot undergo the same phenomena at the same time. Callisto is not involved in this relationship. It is possible that while two of the inner satellites are in transit

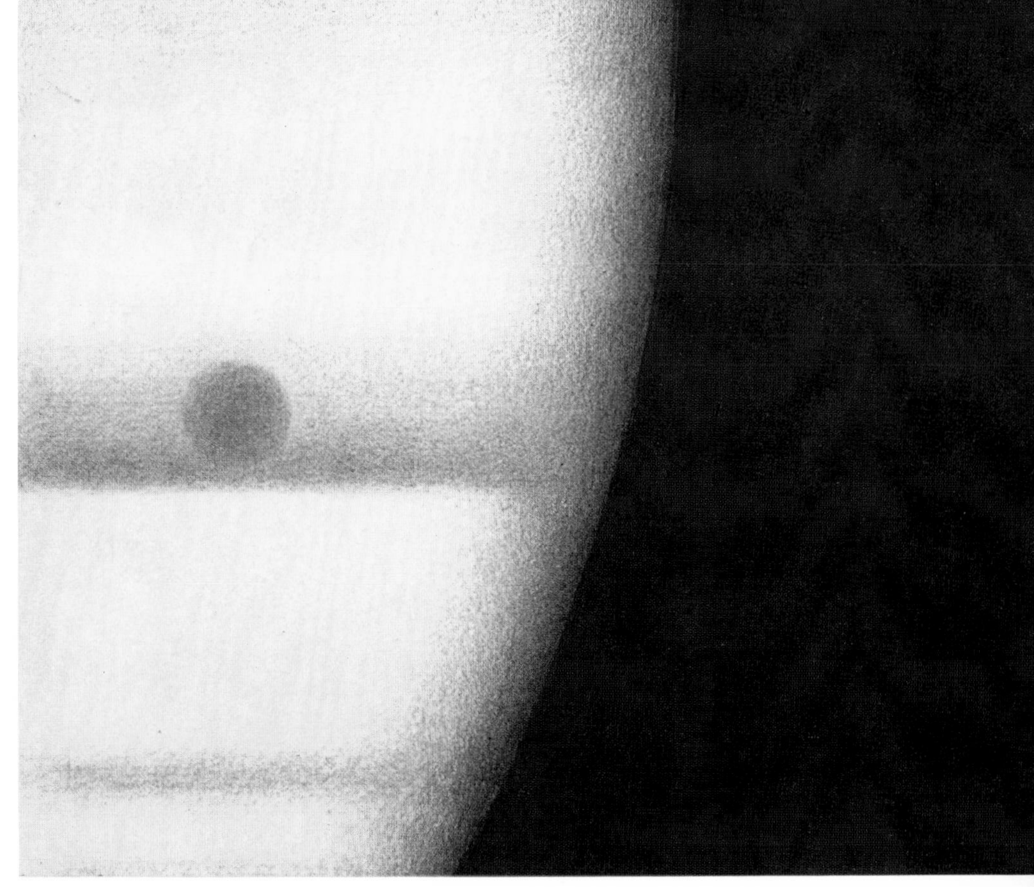

22.VIII.73 22.33–23.29 UT 254mm Refl ×280 The albedo of Satellite IV is so low that it appears to be very dark when seen in transit.

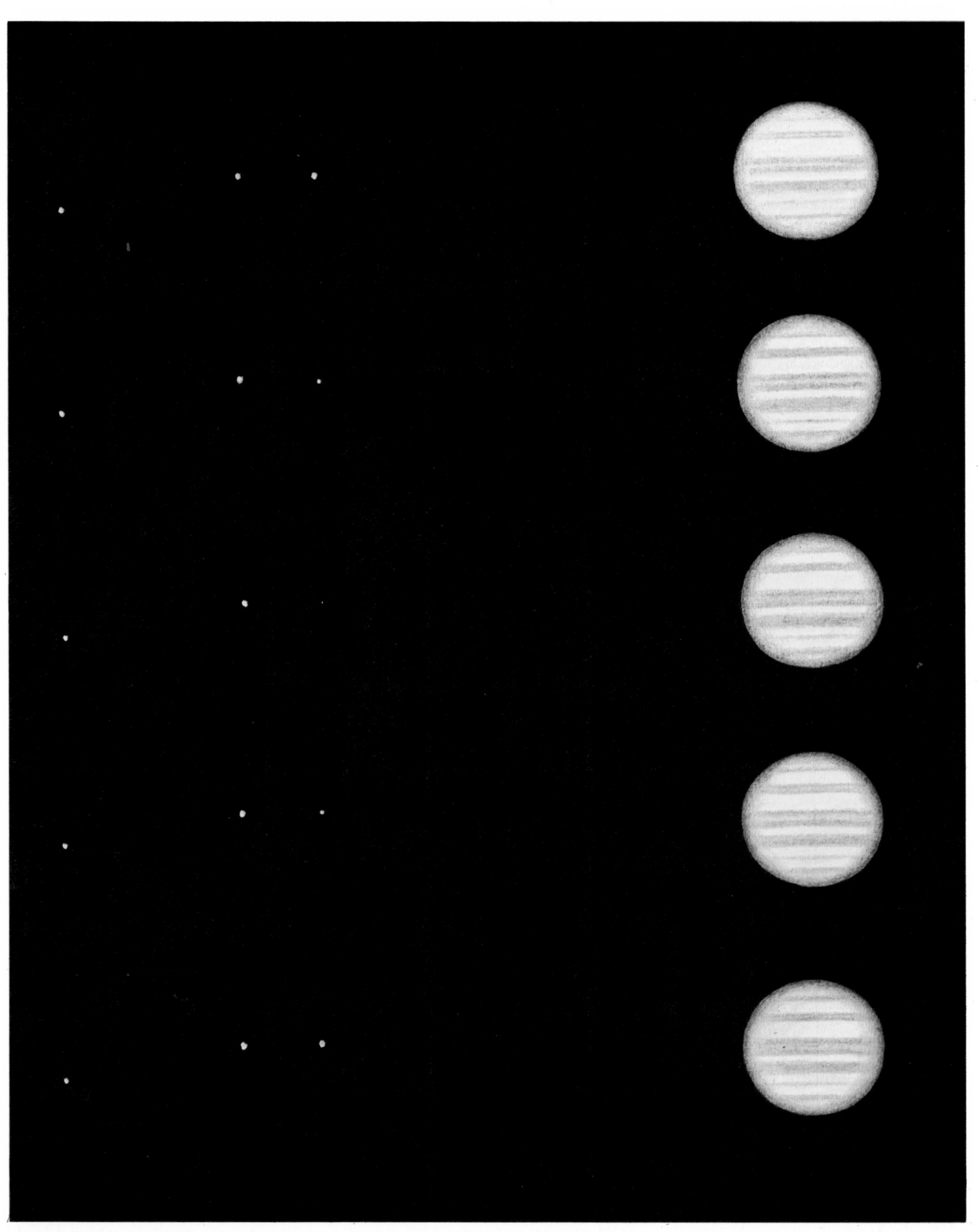

regions showed a mass of confused detail. Belt structure here was difficult to make out because of a multitude of spots and swirls. Along the edges of certain belts detailed cloud swirls and eddies were visible. The cloud flow around the Great Red Spot was of particular interest since it moved in a different direction north of the spot to what it did to the south. Dusky regions preceding and following the Red Spot are caused by interactions of the opposing flows. A clear spiral structure could be seen in the RS itself and in many of the white

Left *23.VIII.73 20.05–20.12 UT 254mm Refl ×48 Eclipse of Satellite II (Europa) by I (Io).* Right *6.IX.73 23.25–23.41 UT 254mm Refl ×280 Occultation of Europa by Io.*

spots and ovals small, dusky, central spots were found that resembled the eyes of Earth-type hurricanes. It seemed, in fact, that these ovals and spots of the temperate regions were actually circular storms.

With all the information amassed so far, astronomers are trying to build up a picture of what Jupiter is really like. It seems that the hydrogen and helium composition of the planet is of similar relative proportions to the Sun. This fact has led to the suggestion that if Jupiter had been a more massive body it may well have become a Sun itself. Since, however, it is not so massive, the temperature at its centre is not sufficient to bring this about. Even so, this temperature is still incredibly high and is, in fact, believed to be up to 30,000°K, or about five times the temperature at the Sun's surface. The temperature at the centre of the Sun is, of course, many millions of degrees.

At the centre

If we could somehow penetrate the atmosphere of Jupiter and travel to its very centre here is probably what we would find. First we would encounter a thin outer atmosphere of hydrogen and helium, followed by high-altitude clouds of ammonia ice crystals. A further cloud layer, dark orange or reddish brown in colour, would then be found which may well contain ice crystals of ammonium hydrosulphide. This substance actually changes colour with exposure to sunlight due to a photochemical reaction and may give rise to the strong colours we often see in the dark cloud belts. Next, a layer of dense water mist is thought to exist. The whole of the planet's atmosphere also contains a large amount of methane. At this point we would be at an approximate depth of 180 kilometres and the atmospheric pressure would be around 100 times the Earth's at sea level. Temperatures here could be about 420°C, which are similar to those found on the surface of Venus.

As we gradually penetrate to a depth of 1,000 kilometres, hydrogen, under great pressure, turns into liquid form. At a still greater depth of more than 20,000 kilometres, the enormous pressures result in this hydrogen assuming a solid metal form so, at the centre of Jupiter, there is a huge sphere of solid metallic hydrogen, more than 90,000 kilometres in diameter. It is possible that at the very centre of this there is a small rocky or iron and silicate core about 12,000 kilometres in diameter. Temperatures here are exceptionally high and pressures are something like 100 million times the atmospheric pressure found here on Earth at sea level.

It is this great heat within the planet that gives rise to the convection of the atmosphere and results in all the turbulence we see. We know that air is rising in the zones and sinking in the belts. Coriolis force (the deflecting effect of the planet's rotation) is negligible at the equator and in this region the belts tend to be well defined. Towards the poles, however, we have the confused detail and circulatory cloud patterns and the Coriolis forces play an important part in the cloud formations of these regions. These forces drive any north- or south-bound currents into strong easterly or westerly winds. In some cases, jet streams occur and evidence of these may, at times, be seen from the Earth. Rapid moving spots associated with outbreaks in the SEB and NTB are thought to be carried along by these jet streams.

The Great Red Spot is simply a giant eddy. In the past, many suggestions have been put forward for what has been considered as one of the greatest mysteries of the Solar System. These theories have ranged from a huge volcano to a great lump of solid helium floating in Jupiter's atmosphere. It seems that Pioneer has given us the answer to the Red Spot problem, although all the clues were there in Earth-based observations. Years ago it was suggested that this feature was a Taylor Column that formed over a mountain or large crater on the surface of Jupiter. As long ago as 1968 observations showed dark spots of the SEBs and STBn circulating around the spot with a period of something like twelve days. This led E. J. Reese to suggest that the spot was a giant vortex. At least the rotary motion of the spot had been recognized by observation.

Most adventurous

Along with the Pioneer results, Earth-based observations have still managed to take a few great steps forward. Observations and photographs of a huge sodium cloud surrounding Io are most remarkable.

As if all this were not enough, during 1977 the Americans launched two more probes to Jupiter. Voyager 2 lifted off first, on August 20, followed by Voyager 1 on September 1. These probes can probably be described as the most adventurous to date. On passing Jupiter they are both designed to continue on to Saturn, and Voyager 2 will, it is hoped, travel on to Uranus.

Observations confirmed

Prior to Voyager 1's close encounter a series of time-lapse pictures were taken. These showed the rotation of the planet apparently speeded-up. More fascinating than this was a longer series which showed the actual different flow patterns and interacting currents within the atmosphere. All that had been derived by painstaking observation in the past could be seen instantly in one brilliant moving illustration. Different motions around the RS clearly showed, together with indications of motion with the RS itself as it appeared to rotate. The difference between the rotation period of systems I and II showed magnificently, and the speeded-up comings and goings of the Galilean satellites were a joy to watch for anyone used to seeing them in the natural slow motion.

An intensive scientific study began on February 26, 1979, and long before actual close encounter some amazing detailed pictures were obtained. With the spacecraft still some 8 million kilometres away, pictures showed the two inner moons, Io and Europa, superimposed over the multi-coloured surface of the planet and clear albedo markings could be seen. Europa displayed a dusky, purple brown, equatorial band and bright whitish poles. Io had distinct reddish poles with bright yellow equatorial regions. These backed up previous Earth-based findings. One picture of Io showed some ellipticity in the satellite's figure. One astronomer, W. H. Pickering, when making measurements of Io's disc in 1892, stated that he had found it to be elliptical. Soon after, he found similar results for the other three major satellites. Pickering's observations were never confirmed and were, in fact, strongly criticized. This picture of Io from Voyager makes one wonder.

Most interesting of all were the cloud patterns of the planet itself. Every colour of the rainbow showed in a magnificent psychedelic pattern, which resembled a mixture of brightly coloured paints floating on oil, mixing together but with individual colours remaining separate. Some parts had the appearance of cream poured on to coffee.

At this time a large STB white oval was passing through conjunction with the RS, and incredibly complex patterns resulted from the interaction of the two features in the part of the STB that followed. The circulatory pattern of the RS showed clearly, although things seemed calm at its very centre. Also, a similar circulatory structure in many of the white spots and the STB oval could be seen. Preceding both the RS and STB oval, the belts were neither highly coloured nor detailed, the belts appearing red-brown, the zones yellow ochre to white. Following the two features, the belts and zones of the STB, STrZ and SEB were in turmoil. Swirling cloud eddies along the STBn and huge bright swirling clouds on the STB showed clockwise and anti-clockwise motion.

Jupiter's ring

During a programme of Earth-based observation carried out during the months before Voyager's arrival, amateur and professional astronomers had studied the life history of certain features on the planet and had gained a clear knowledge of the drift rate of those features. The author's observations at the time of the Voyager approach indicated that the RS had only a slight drift in the direction of the planet's rotation. We also knew, of course, that the STB white oval was passing the RS. It appears from this that the turbulence of the STB and SEB forms in the wake of these features. A combination of the Voyager and Earth-based observation will tell us much about the interacting currents involved here. This illustrates once more the valuable part the amateur can play even in the space age.

Actual close encounter for Voyager 1 occurred on March 5, 1979, when the spacecraft passed within 278,000 kilometres of Jupiter's cloud tops. It also passed within 440,000 kilometres of Amalthea, 25,000 kilometres of Io, 750,000 kilometres of Europa, 130,000 kilometres of Ganymede and 130,000 kilometres of Callisto. The highly detailed pictures were well beyond expectations. Very dark streaks, looking like holes through Jupiter's clouds,

16.1.77 20.46 UT 419mm Refl ×300 Two dents in the limb caused by satellite shadows.

were photographed. Strongly contrasting vivid reds and bright white in the Red Spot region showed clearly.

One surprising thing was the discovery of a ring around the planet at a distance of 55,000 kilometres, which is closer than the innermost moon. This is a very tenuous feature, 30 kilometres thick and 8,000 kilometres wide, but it now means that Jupiter is the third gas planet to display such a phenomenon. This ring is thought to be composed of rocky debris and it is not known if it was formed at the time the Solar System was formed or if it is the result of a small moon torn apart by tidal forces. Also, because it was photographed edge on, it is not known if there are any divisions in the ring.

Detailed pictures of the satellites were obtained. Dark markings on the two outer moons, Ganymede and Callisto, are thought to be the result of some form of vulcanism. These two moons are composed largely of ice and rock. Io shows a very smooth surface which lacks the large number of impact craters displayed by the other moons. There is extensive vulcanism on this satellite and not less than six active volcanoes were photographed.

Favourite planet

One eruption was sending plumes of material to a height of 100 kilometres. A large volcano, 170 kilometres in diameter, showed dark, apparently basaltic, lava flows at least 100 kilometres long by 15 kilometres wide. The smoothness of Io's surface is a result of this recent vulcanism and there is a surface deposit of sulphur and salt which gives the satellite its yellow and vivid orange colours. Io, it seems, has an active molten interior similar to the Earth's.

Most fantastic were the lightning flashes on Jupiter seen by Voyager, from a distance of 6 million kilometres, as it left the planet for Saturn. Some flashes were about 32,000 kilometres long and these are taken to indicate that the planet has a larger magnetic field than originally thought. Intense Auroral activity has also been recorded on both the day and night side of the planet.

Jupiter must be considered as the major interest in planetary astronomy. It is certainly the finest subject for amateur observation. Its ever-changing cloud patterns, incredible Red Spot and bright moons make it first choice for most planetary observers. The new findings by Voyager will only add to its already enormous fascination and interest for man.

Saturn

After the violent turbulence of Jupiter it may come as something of a relief to turn to the apparently serene beauty of the magnificent ringed planet Saturn. To most astronomers Saturn, though very beautiful, lacks the fascination of Mars and Jupiter because it does not display their wealth of detail. However, it would be a mistake to think that the atmosphere of Saturn is not subject to violent currents of the type clearly seen on Jupiter—as will become apparent during the course of this chapter.

Saturn is certainly the most attractive and memorable sight for anyone taking their first look through a telescope. The author, since 1959, has viewed this planet on many occasions, sometimes through large telescopes, and has never failed to be struck by its incredible beauty. But no view has ever been as pleasing as the first, with only a 75-mm refractor, which made an ever-lasting impression.

Saturn is the second gas planet in order of distance from the Sun and the second largest of the planets. It resembles Jupiter in many ways. Its mean distance from the Sun is 1,427 million kilometres, varying between 1,347 million kilometres at perihelion and 1,507 million kilometres at aphelion. The equatorial diameter of the planet is 120,000 kilometres, which is about nine and a half times the diameter of the Earth. The volume of Saturn is just over 750 times that of Earth, but its mass is only 95 times as great as ours. Even though you could pack no less than 750 Earths on to a globe the size of Saturn you would only need 95 Earths to equal its weight. From this we know that Saturn has a very low density. In fact, it has the lowest density of all the major planets, amounting to no more than 0.7 times the density of water. Thus, a sphere of water of the same size as Saturn would weigh much heavier than Saturn itself.

Bulging equator

Like Jupiter, Saturn has a rapid rotation, turning on its axis once every 10 h 14 m and resulting in a very marked polar flattening. Centrifugal force makes Saturn bulge at the equator so that its 107,000-kilometre polar diameter is 12,500 kilometres less than its diameter at the equator. This difference is as large as the equatorial diameter of the Earth, and the oblateness is a feature clearly visible, even in very small telescopes.

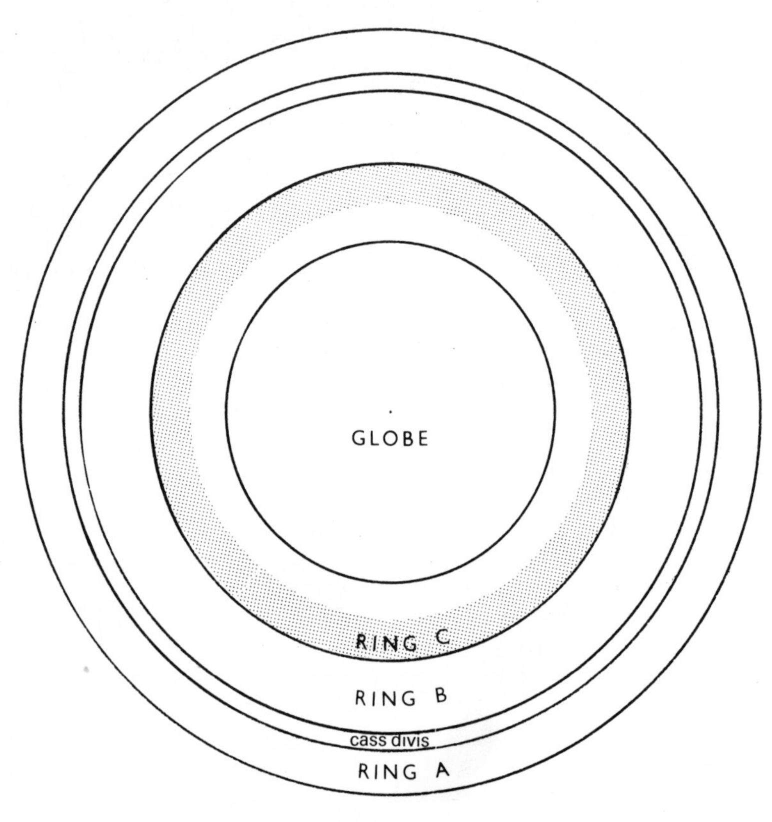

Nomenclature of Saturn's rings.

Even though it is a great distance from us the large diameter of Saturn yields a reasonably-sized telescopic image. At very favourable oppositions, it has an angular diameter at the equator of over 19″. The angular size will, of course, change with the varying distance of the Earth from the planet. At the most favourable oppositions its angular diameter approaches 21″ whereas at conjunction it may only be 14″. However, at average oppositions a magnification slightly less than × 100 will give a size for the globe comparable to the Moon as seen without optical aid.

It is the ring system of Saturn that lends it its beauty. The dimensions of the rings are staggering. Their outer diameter is no less than 272,000 kilometres, well over two-thirds of the distance between Earth and the Moon, and their overall width is in the region of 65,000 kilometres. The 'system' comprises three concentric rings, including an outer ring with a width of 16,000 kilometres (Ring A) and a brilliant inner ring (Ring B), 28,000 kilometres in width. Separating these two rings is a 4,000-kilometre gap called Cassini's Division. Finally, there is a faint inner ghost-like ring that has a width of 17,000 kilometres, known as the Crepe Ring, or Ring C. Despite their enormous surface area it is believed that the rings are only a few kilometres thick. Nevertheless, they are very striking when viewed from the Earth. At opposition, the full angular extent of the rings may be as much as 45″ so that they subtend an angle roughly comparable to the equatorial diameter of Jupiter.

Belts and zones

Saturn takes over 29½ years to orbit the Sun once. It has an axis of rotation tilted 26.7° to its orbital plane and therefore experiences seasons like the Earth and Mars. These are of course quite long, lasting on average 7 years each. The ring system lies in the plane of the planet's equator and so the axial tilt means that we see not only the globe of the planet at varying degrees of tilt but the ring system as well. At their widest opening, the rings may be tilted by as much as 26° to the Earth's line of sight. They

14.XI.75 00.55 UT 254mm Refl ×280 The tilt of the planet allows a most favourable view of the rings and the southern hemisphere. The black Cassini Division and the globe's shadow on the rings at upper left can be seen here quite clearly.

are, however, presented edgewise on to the Earth at times when the Earth is passing through the ring plane. More will be said about these aspects later.

The globe of Saturn resembles closely that of Jupiter in that it is marked with belts and zones that run parallel to its equator. It is generally assumed that the reason for their existence is the same as for those on Jupiter. Convection cells that circle the planet with air rising in the zones and sinking in the belts are forced into their belt-like appearance by the planet's rapid rotation. For the most part, belts do not show the kind of detail exhibited by the belts of Jupiter but the basic pattern is the same and the nomenclature similar to that given in the chapter on Jupiter.

The equatorial region of the planet is marked by a wide bright Equatorial Zone (EZ), which is occasionally marked near its centre by a faint dusky Equatorial Band (EB). This zone tends to be a little duller to the north than to the south. Flanking the zone to the north and south are the dark Northern and Southern Equatorial Belts (NEB and SEB). They are similar to their counterparts on Jupiter but are less active and are generally of a higher latitude—almost equivalent to what would be Jupiter's tropical regions.

Eclipse and occultation

Moving further towards the poles we encounter the Tropical Zones and then the Temperate Zones and Belts, each having the prefix S or N depending on which hemisphere they are in. The number of times that either prefix is used denotes the approximate latitude, that is, the more the prefix is used the higher the latitude of the belt. Thus, NNTB is closer to the pole than NTB. The polar regions themselves are dusky and are sometimes also marked by small dark polar caps.

The three rings of Saturn both cross in front of, and pass behind, the globe, leading to mutual eclipse and occultation phenomena. The black shadow of the globe will normally be seen falling on to the rings as they pass behind the ball of the planet. It will be cast either on the leading part of the ring, or preceding ansa, prior to opposition or on the following ansa after opposition. At the time of actual opposition the shadow falls directly behind the ball of the planet and is, to all intents and purposes, hidden by it.

The shadow of the ring on the globe will be seen accompanying the ring itself where it crosses in front of the ball. This shadow may fall either to the north or to the south of the ring, depending on the Sun–Earth–Saturn angle at the time of observation. Sometimes it is completely hidden by the ring and is not therefore visible.

As with Jupiter, it is the upper layer of a very dense atmosphere that we see and not the true surface of Saturn.

If it were possible to view the rings of Saturn from close up this is possibly what they would look like. The rings are probably composed of ice crystals with some rock fragments. The globe of Saturn can be seen clearly on the left and one of its inner moons in the distance.

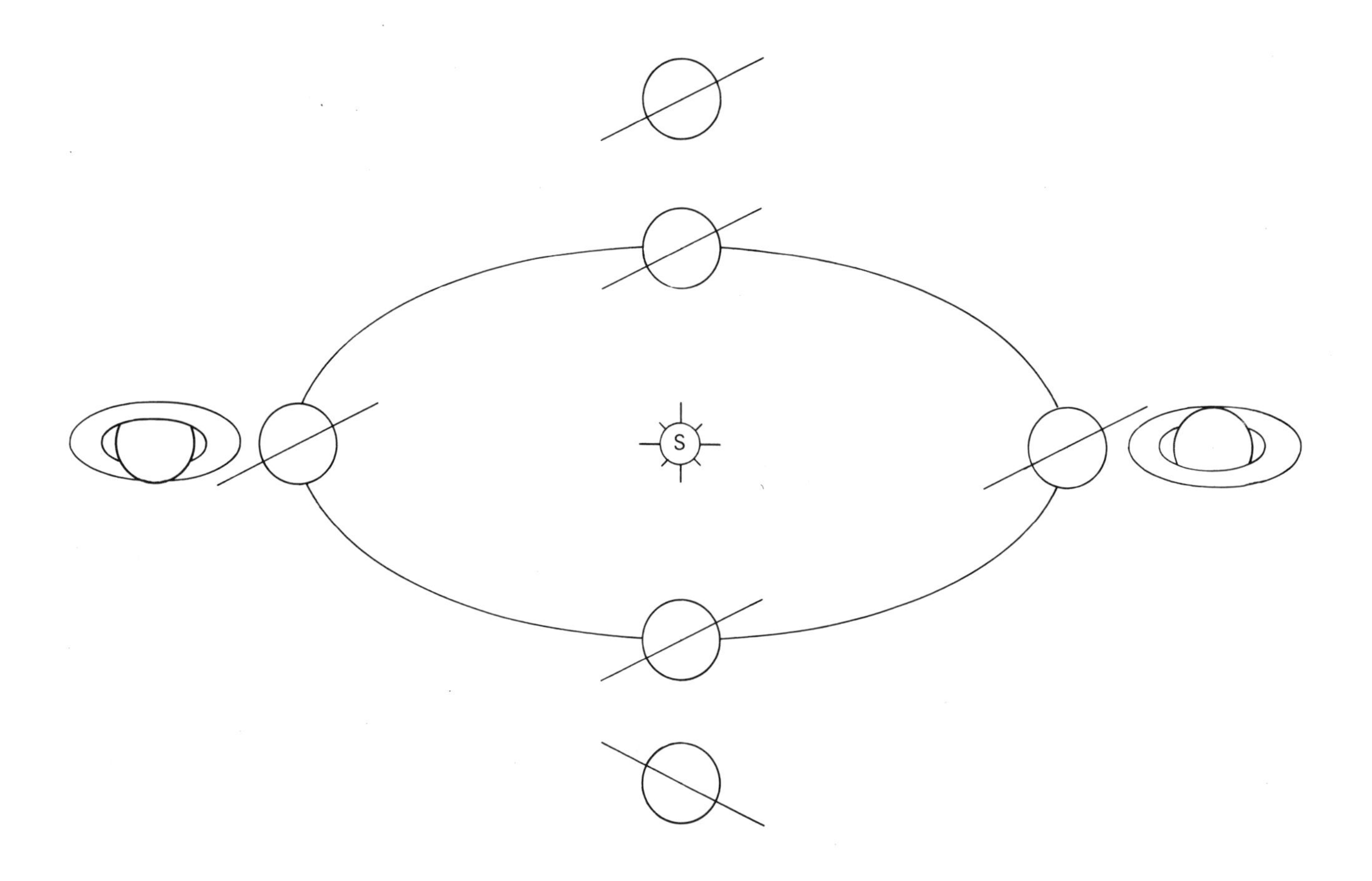

Different aspects of Saturn's rings showing why they appear either edge-on or wide open.

There is plenty that the amateur can do by way of observing the globe of the planet. One of the most important tasks for the amateur is to estimate the intensity of the various features seen, including the rings. A recognized scale of intensity has been adopted, which gives a value of 10 to the black surrounding sky and 1 to the bright outer part of Ring B. For such purposes, this part of the ring is regarded as having a constant light value, though this is not strictly true. Any feature displayed by the planet is estimated in relation to these two standards, the idea being to establish any seasonal variation that the markings might display. This would only be slight and would be spread over many years, so continued observation is essential. Usually the darker belts have an intensity ranging between 5 and 6 while the bright Equatorial Zone seems to alternate between 1 and 2, which is similar to the intensity of Ring B. There are times, however, when intensities move beyond these limits. If any feature appears brighter than Ring B then it will be given a value of 0, or even a minus value. Such instances are, at best, very rare.

Longitude and latitude

Another task for the observer, which is of exceptional importance, is the determination of longitude for any bright or dark patch or spot seen on the belts or zones of the planet's globe. Features of this sort are rather rare on Saturn but they are not as rare as many astronomers tend to think. The problem is that such features usually have a small intensity difference from their surroundings and are thus difficult to detect. They further tend to have softened outlines, making it awkward to determine their preceding and following ends, and are, moreover, often very short-lived.

The chances of seeing the same feature more than once, even over an extended period, are slight, making it difficult to obtain a reliable drift rate for any individual feature. It is also difficult to determine the longitude accurately, because of the nebulous nature of most of the features seen. For this reason, there is often a large margin of error between observations. Furthermore, important features are often missed entirely, because of the paucity of observers following the planet, and altogether there is a depressing lack of knowledge about the various currents within the planet's atmosphere.

It is interesting and useful to note that seven rotations of Saturn's equatorial region are equal to three rotations of the Earth, less 22 minutes. Thus, any feature of interest timed across the CM will, after a period of three days, again be in the vicinity of the CM 22 minutes earlier than the original timing. This is particularly useful for speedy confirmation of any feature.

The determination of the latitude of the various belts is also possible and is a very valuable type of observation. Unfortunately, the usual method of deriving the latitude of Saturnian features is rather laborious and many observers prefer not to bother. Micrometric measurements at the telescope are best, of course, but accurate results can be obtained from measurements of drawings. Providing there are enough observers engaged in the work, an average taken from their observations can be quite accurate. As in the case of the intensity estimates the point of the exercise is to try to establish any seasonal variation in the position of a belt or in its width. It so happens that there is not much variation in the latitude of the belts and nothing that can as yet be put down to seasonal variation. This is to be expected, however, when the formation of the belts is a result of heat from within the planet itself and not heat from the Sun. The method of calculating Saturnicentric latitudes is as follows:

Obtain B, the Saturnicentric declination of Earth, for the date of observation. (This can be found in most astronomical handbooks.)

From $\tan B' = 1.12 \tan B$, find angle B'.

Now, from $\sin (b'-B') = y/r$, find $(b'-B')$, and hence b'.

Finally, from $\tan C = (\tan b')/(1.12)$, find C.

C is then the required Saturnicentric latitude.

The measurements used on the drawing are the polar radius (r), expressed in millimetres, and the distance (y) of the point whose latitude is required, from the centre of the disc of your drawing, again expressed in millimetres and measured along the CM.

Measurements of any dark belt should be made, in particular its northern and southern edges. Also, the inner edge of ring C, where it crosses the globe (Cm) is of value.

Size of telescope

As for the type of instrument needed to carry out useful work on Saturn, it is generally accepted that nothing less than a 125-mm refractor should be used. It is true, however, that at times certain important features have been seen with telescopes of only 75-mm aperture. Normally, a larger telescope is desirable and the best size may be between 250 mm and 300 mm for reflectors or 150 mm for a refractor. An attempt may, of course, be made to better these and something around 600 mm would obviously give the best results. The planet actually takes magnification well and the maximum recommended magnification for a given telescope can often be used. In general, the brightness of the image is all important when looking for any slight variation in intensity. The author has found $\times 250$–$\times 300$ to be ideal with a 420-mm reflector.

Once equipped for observation, the observer should be aware of certain illusions—usually resulting from the very nature of a planet with rings—which may cause him to draw erroneous conclusions about Saturn.

Illusory effects

The most notorious of all these illusions is known as the 'Herschel square shoulder effect', which was, not surprisingly, first noticed by the famous astronomer Sir William Herschel at the end of the 18th Century. He made measurements of the figure of Saturn's globe and found that not only did it appear flattened at the poles but also at the equator. This seemed to give the globe a squared shape. Herschel has, to this day, commanded great respect, and it was not until 50 years after his 'discovery' that the astronomical world acknowledged that Herschel had been mistaken and that Saturn's globe was a true elliptic spheroid.

How was it that Herschel had been fooled? It could have been a combination of effects. Most of his observations were made with Saturn low down in the sky as seen from Britain, so that the seeing was probably poor. Also, particularly dark polar areas, with bright surrounding zones could have created the impression of a slight bulge on the limbs near the poles and a flattening of the polar limbs themselves. This, together with the rings surrounding the globe, might have given an apparently squared shape. Still it was an error, and if so great an observer as Herschel could be mistaken, then others less skilful most certainly can.

Another unusual effect concerns the shadow of the globe on the ring. This is the appearance of a peak at the point where the shadow crosses Cassini's Division and is most noticeable when the rings are wide open. At such times, it gives rise to what is called the 'peak shadow effect', which may be seen on some of the drawings presented here. At first, it was taken to mean that the rings must vary in thickness throughout their width. However, we now know that the rings are very thin, and the sort of thickness variation required to produce such an effect is not possible. The variation in thickness would in fact need to be much greater than the true thickness of the rings.

The real reason for the effect is simply that telescopes and seeing conditions are far from perfect. The phenomenon is in some ways akin to the black drop effect on Venus. A blending together of the black shadow and the Cassini Division, coupled with irradiation making the bright ring components eat into the dark sky all brings about a very obvious illusion that shows up even on photographs.

Yet another illusion involves the apparent shadow of the ring itself. This shadow has been seen on both sides of the globe at the time of opposition, when in fact it should not have been visible. This is not so much an illusion as the pure effect of very darkened limbs crossing the bright ring.

Still connected with the ring shadow is an effect called 'Terby's white spot'. This was first seen in 1889 and has been regularly observed since. It is the appearance of a brilliant spot on the rings, adjacent to the ring shadow. At times, it is very obvious although it results, in fact, from pure contrast and never shows on photographs.

Generally inactive

Although the atmosphere of Saturn appears generally inactive, the observer can, at times, expect sudden violent and spectacular outbreaks of activity. Changes are, for the most part, however, restricted to the gradual darkening or fading of the belts and the occasional brightening of the zones. These changes tend to spread over a few weeks or even months. Sometimes no change at all will occur throughout a whole apparition.

It seems that when a hemisphere that has been hidden from the Sun for a number of years first becomes visible, it has a different appearance and colour from the hemisphere that has been exposed to the Sun for some years. Gradually, with exposure to the

Sun, the new hemisphere starts to resemble the old, Sun-turned hemisphere more and more. The colours become brighter and the contrast between the belts becomes more marked. This may be the result of some photochemical reaction in the materials that constitute the atmosphere of Saturn. But only prolonged and diligent observation will establish a definite pattern.

Spots on Saturn, either light or dark, are rather infrequent. If anything, the dark spots are the more common but they are less spectacular than bright spots. Observations of these spots are very important for acquiring a good knowledge of currents within the atmosphere. Yet, strangely, the first estimating of the planet's rotation period was made by William Herschel in the complete absence of prominent spots. He used, instead, the slight changes in the thickness and intensity of the belts. His rotation period, given as 10 h 16 m plus or minus 2 minutes, is very close to the accepted period of the equatorial region we have today. This is indeed proof of Herschel's incredible skill.

The first recorded well-defined bright spot occurred during the apparition of 1876. It was discovered on December 7 by Asaph Hall with the Washington 650-mm refractor (the telescope with which he discovered the moons of Mars only 12 months later). A bright white spot was situated on the Equatorial Zone and it lasted a number of weeks. From this, a rotation period of 10 h 14 m 23 s was eventually derived.

Following upon this success, during the 1890s amateur astronomers, and most notably A. Stanley Williams, recorded many light and dark spots in the equatorial regions all of which agreed very closely with the 10 h 14 m-rotation period. These observations were, however, strongly criticized by the professional astronomer E. E. Barnard, who failed to see any spots of the type seen by amateurs even though he was using the 900-mm Lick refractor. Then, on June 15, 1903, Barnard himself found a bright white spot, while using the 1-metre Yerkes refractor. This spot was one of a number seen during the ensuing weeks. But unlike those seen by Hall and Williams, these spots were at a much higher latitude and were situated in the Northern Temperate Zone. Preliminary calculations for the rotation periods of the spots gave 10 h 39 m, which seemed very long compared to other rotation periods for equatorial spots. The final average period for these NTZ spots came out at 10 h 38 m and it was quite certain that there was no error. This led A. Stanley Williams to conclude that Saturn has a great equatorial current which rotates much more rapidly than the temperate and polar regions.

The difference between Saturn's two systems of rotation is much greater than that found on Jupiter. The material of the atmosphere at Saturn's equator travels at 1,300 kph faster than the material at the poles, compared with a difference of 320 kph in the case of Jupiter. From this, it would seem that the atmosphere of Saturn is subject to a much greater mixing than the atmosphere of Jupiter, which may be why marked individual features on Saturn are very short-lived.

13.I.78 00.15 UT 419mm Refl ×248 Although the atmosphere of Saturn is much less turbulent than that of Jupiter, occasional spectacular features are recorded. Shown here are equatorial white spots like the one seen in 1933.

Famous white spot

The most famous feature of Saturn's globe is the bright white spot of 1933, discovered by the comedian Will Hay on August 3 of that year. Its fame resulted not only from its discoverer but from its unusual prominence for a Saturnian feature. It was also remarkably long-lived by Saturn standards and was followed for at least six weeks. This has taught us a great deal about the motions of Saturn's atmosphere.

When it first became visible the spot was oval in shape and quite well-defined, with dimensions estimated as something like those of the Great Red Spot on Jupiter. However, the preceding end of the spot soon began to appear drawn-out and ill-defined. Gradually the spot lengthened until after a few weeks the preceding end became difficult to discern and transit timings both of this end and of the centre of the spot became harder and harder to make. Finally, the following

end of the spot was the only part remaining sharply defined, but it was from timings of this part that the most accurate rotation period was determined. (It should be mentioned that this lengthening of the spot was a characteristic of the previous spots observed by Hall and Barnard.)

The period of Hay's spot averaged out at 10 h 14 m 12 s. At first, the spot tended to have a slow rotation of 10 h 16 m but during its lifetime it accelerated to 10 h 13 m. It seemed from this that the material of the spot rose to a high level in Saturn's atmosphere and perhaps encountered a fast overriding wind, which brought about this acceleration. Fortunately, the source of the spot must have been long-lived, and so, although it spread rapidly in the direction of the planet's rotation, the source feeding the spot resulted in a long-lived, clearly defined, following end. The spot must have been at a higher level than the surrounding material, because photographs taken in ultra violet light showed the spot but not the surrounding bright zone material. The whole situation was afterwards complicated by the appearance of more spots and by an apparent splitting of the main one, all of which resulted in a marked brightening of the EZ.

Notable outbreak

The next notable bright spot outbreak occurred during the 1960 apparition and was discovered on March 31 by the South African amateur, J. Botham. This particular feature occurred at an even higher latitude than Barnard's spot of 1903, being situated in the North North Temperate Zone. When discovered, its preceding end was diffuse but the following end was quite sharp. The feature was fairly large, taking over one hour to cross the CM. Subsequent

Above left *11.X.69 04.20 UT 450mm Refl ×350 Faint dusky filaments on the Equatorial Zone.* Left *11.X.73 04.30 UT 254mm Refl ×380 Dark spots on the South Equatorial Belt.* Above right *15.I.76 01.00 UT 254mm Refl ×300 An unusually light South Polar Region.* Right *27.IV.76 21.30 UT 254mm Refl ×280 Changes in atmospheric markings (cf. above left).*

accurate timings of the preceding end and centre of the spot became increasingly difficult because of a rapid spreading of the spot material in the preceding direction. In less than two weeks, the spot covered an appreciable area of the NNTZ and the following end was again the only part to remain clearly defined. The final rotation derived for this feature ranged between 10 h 39 m and 10 h 40 m. This result was in keeping with the spot seen by Barnard in 1903 and confirmed the fact that there is a great equatorial current.

Patterns of activity

Further spots in this region were observed independently by a number of astronomers. Of particular note was a feature seen by A. Dollfus, for which he derived a rotation period of 10 h 39 m. It is not entirely certain that the feature seen by Dollfus was the same as the one seen by Botham since Dollfus picked up his spot towards the end of April, by which time Botham's seems to have faded. Whether or not it was the same one, the rotation periods derived were very similar. The appearance of the spot was followed by a considerable amount of activity in the region, resulting in a widening and brightening of the NNTZ. At the same time Dollfus managed to obtain polarimetric measures of the bright region and discovered a much higher level than for the general zone material.

Although it is difficult to generalize with this type of feature, the few bright spots that have appeared tend to follow a certain pattern. When first formed they are bright and well-defined, but within a few days they spread out rapidly in a preceding direction, resulting in an eventual brightening of the zone where the spot first appeared. These spots seem to be high-altitude clouds that become involved in a high-speed wind dwelling in the upper regions of Saturn's atmosphere and, providing the source of the spot continues, the following end of the spot will remain well-defined.

There have been one or two recent outbreaks of a less spectacular nature. A bright spot on the EZ was seen in 1962. In 1968 three large ill-defined light patches appeared on the EZ. This was followed by a similar appearance of three white EZ spots in 1972

and then, in 1976, a small bright spot on the SSTZ was recorded.

Perhaps the best outbreak of recent times occurred in 1978. A small bright spot was observed on the EZ by astronomers in Italy and England. The derived rotation period of this feature was 10 h 18 m, which seems a little long but is not beyond the accepted limits.

The darker spots

More common than the bright spots are small dark spots that appear on the belts. These are not usually very prominent and are most often seen along the southern edge of the NEB and the northern edge of the SEB. An important series of observations of this type of feature, was carried out during the 1890s and, once again, A. Stanley Williams was very much involved. The region on which they appeared was the NEB and rotation periods derived ranged between 10 h 12.5 m and 10 h 15 m.

A more recent outbreak occurred in 1947, when observers in Britain and America followed a number of dark SEB spots. The general rotation period derived for these ranged between 10 h 14 m and 10 h 16 m. One interesting fact was that the spots with the slower rotation seemed closer to the equator, which is surprising when you consider the general slow rotation of high-latitude regions. Another interesting feature was the appearance of one or two dark patches that had very rapid rotation periods of the order of 10 h 9 m. One even attained a rotation period of 9 h 55 m, which is equivalent to System II on Jupiter. It is evident, from these periods at least, that the Equatorial Belts are part of the Equatorial Current but it also seems possible that strong jet streams might exist in these regions and that the different rotations might result from the varying altitude of the features observed in the atmosphere of Saturn. Clearly, more observations are needed before we can draw any proper conclusion.

In 1973, a further quite active outbreak of spots occurred on the SEB, as shown in some of the drawings presented here. Again, the rotation derived for these was quite rapid, at something like 10 h 13 m.

The globe of Saturn is interesting enough but the planet has something else, of course, that can only be classed as extraordinary—its system of rings. It is no longer correct to say that this ring system is unique, since rings have now been found around Uranus and Jupiter. Neither of the latter are readily observable, however, so from this point of view Saturn does offer the visual observer a unique opportunity to witness clearly a magnificent phenomenon. The story of the discovery of the rings and the gradual understanding of their true nature is fascinating and warrants being told in a little detail.

When Galileo took the first ever glimpse of this planet through a telescope he was surprised to see not one single disc but three—one large disc with a small one on either side. At first, he thought these were like the moons of Jupiter, which he had discovered shortly before but, on seeing that the positions of the small bodies did not change in relation to the larger body, he realized that this could not be. More puzzling still was the fact that when he again looked at Saturn, in 1612, the small discs had disappeared. He had no idea what was happening and a number of his critics at the time suggested that what he saw might be just a figment of his imagination. But a short time later the two small discs reappeared and Galileo just could not understand why this should be so.

The ring system

The true nature of the phenomenon remained unsolved until 1656, when Christiaan Huygens realized that the planet was surrounded by a thin flat ring. Further study of this ring brought about the discovery by J. D. Cassini, in 1675, of a division separating the ring into two main components. We know now the explanation of the event that was so puzzling to Galileo. Every so often the rings of Saturn appear edge-on to the Earth and, because of their thinness, they are invisible in all but the largest of modern telescopes. The quality of Galileo's telescope was such that he lost the ring for a considerable period of time. He would also only have seen the broad extremities of the ring, and these would, to him have taken on the appearance of small detached bodies. It is unfortunate that when Galileo made his first observations, the rings were just approaching their edgewise appearance. Otherwise he might well have realized what they were.

11.X.73 01.20 UT 254mm Refl ×250 The rings, seen at their widest opening as viewed from Earth, are most impressive.

There are many things that an observer can look for in connection with the rings. Even a brief glimpse will show the main components together with the narrow black Cassini Division but closer inspection will show much finer detail. The inner part of

Ring A is brighter than the outer part, while the outer part of Ring B is brighter than its inner part. In moments of exceptional seeing, even more detail can be made out. Many observers have reported what they take to be minor divisions in the ring system that are much narrower than the main Cassini Division. These are usually seen separating regions of different intensity. To discuss these fully, a little more of the history behind the observation of the rings is needed.

Until the middle of the 19th Century many astronomers considered the rings to be solid. William Herschel had shown that the Cassini Division was a true gap in the rings by measuring its position on both faces of the ring and finding that the measurements matched. But the solid ring theory still held. It was only the discovery of the Crepe Ring by W. C. Bond in 1850 that really put an end to this theory. Soon afterwards, it was noticed that the globe of Saturn could be seen through this ring, which meant that it was to a certain extent transparent. This would be very unlikely if the ring were solid.

In 1857 the Scottish physicist and astronomer James Clerk Maxwell put forward a theory that the rings were not in fact solid but composed of millions and millions of tiny particles of dust which orbit the planet like tiny satellites. This fits the evidence perfectly, and is the generally accepted theory to this day. The way in which the ring, or rings, were formed is still in question. Some theories suggest that it may have been a small satellite that approached too close to Saturn and was torn apart by the enormous tidal forces. Subsequent collisions be-

25.X.66 20.20 UT 225mm Refl ×274 The rings can become invisible when edge-on to the Earth. Note the small satellite which is orbiting the planet in the same plane as the rings.

tween the particles then resulted in the belt of fine particles we now see as the ring. Another theory asserts that the material of the ring might never have formed into a satellite and has therefore never been anything other than what it is now. Both theories have their supporters and, of the two, the author prefers the latter. The composition of the ring seems to be largely ice crystals and rock but this does not support the one theory more than the other. It seems, in fact, that most of the inner satellites of Saturn are composed, in the main, of ice and rock.

Even before Clerk Maxwell's theory astronomers realized that a solid ring could not be stable. Theorists suggested that it might be stable if it were composed of many narrow concentric rings that were separated from each other by narrow gaps. Astronomers started to look for such gaps and thus began the regular observation of minor divisions. It seems that they were sought and discovered out of the need to prove a point.

Between the rings

After the rings had been found to be composed of particles, other reasons for the existence of the minor divisions were necessary, which brings us on to yet another theory. In 1866, D. Kirkwood suggested a reason for the fact that parts of the asteroid belt around the Sun appear clear of minor planets. He pointed out that any asteroid whose orbital period is a simple fraction of Jupiter's orbital period is so perturbed by the large planet that it shifts into a more stable orbit. Jupiter, in this way, sweeps parts of the asteroid belt clear, forming what are now called the 'Kirkwood Gaps'.

In 1867, Kirkwood extended his theory to account for the Cassini Division in Saturn's ring. This, he claimed, was caused by the perturbing effects of the inner satellites. Ring particles in this region are shifted into less-affected orbits, much like some of the asteroids. Kirkwood developed his theory further to account for another fairly clear division, the so-called Enckes Division, which had been generally accepted.

The field was now wide open for a fresh attempt at the discovery of minor divisions, and certain astronomers took full advantage of this. One was Percival Lowell (of the Martian Canal fame). Lowell was observing Saturn during 1909 when he noticed three minor divisions in Ring A. He observed the rings again in 1915 when they were wide open and was able to detect no less than 10 minor divisions in the rings, 3 in A and 7 in B. He extended the theory of Kirkwood to account for all of his divisions and did so quite effectively. In recent times the observations of Lowell have been strongly criticized, but mainly for his Martian canals, which we now know do not exist.

One cannot therefore help wondering about his minor ring divisions.

Many amateur astronomers record the minor divisions regularly but professional astronomers using large telescopes often fail to see them and the question of their existence remains open. There are many things that could account for their appearance. Contrast or poor seeing are just two possibilities. The best series of observations carried out on the ring markings were made by B. Lyot during the 1940s, using the 600-mm refractor of the Pic-Du-Midi observatory in France. This telescope is ideally suited to this type of work. The chart he drew up shows a number of light minima in the rings, but little in the way of true or fine divisions. His work was later supported by the photometric studies of A. Dollfus during the apparitions of 1957 and 1958. The 1979 Pioneer encounter did not help in this respect, but the Voyager encounters of 1980 and 1981 may solve the problem. Until then the controversy of the divisions of Saturn's system of rings will continue.

Translucent

After the discovery was made that Ring C was translucent, there was a great deal of speculation about the possibility of the other bright rings also being translucent. If the rings were composed of particles then, quite possibly, a certain amount of light might pass through them. It was a difficult problem and the fact that the shadow of the rings on the ball of the planet always appeared so black indicated that very little light did, in fact, pass through them. The only way that their translucency could be proved was to watch them occult a bright star. It was a long while before such an event was to occur but when it did it was two amateur astronomers who conducted the observation.

On February 9, 1917, M. A. Ainslie and J. Knight reported the occurrence quite independently of each other. Taken together, their observations covered the whole event. Ring A and Cassini's Division passed over quite a bright star of 7th magnitude. The star was undiminished in brightness when viewed through the division, proving that the division was virtually devoid of ring particles. The most interesting effect was that the star remained faintly visible throughout its passage behind Ring A. This showed that Ring A was translucent. Unfortunately, the star did not pass behind Ring B. The observation was superb, nevertheless, and demonstrates just what an amateur can achieve, even with moderate-sized telescopes. Ainslie used a 230-mm reflector and Knight only a 125-mm refractor.

A further occultation of a star by the rings occurred in 1920. This time Ring B was involved. Another star of 7th magnitude was reported as being visible through the ring, indicating that B is also translucent. The rings, however, were at the time very close to being edge-on, and the observation was a difficult one. There have been other recorded occultations of stars by the rings but in most cases the stars involved were not nearly so bright.

The amateur really can make a useful contribution in this type of observation. Each night that an observer views Saturn a thorough search of the sky should be made in the direction of the planet's movement. If there is a star that is likely to be occulted the fact should be reported immediately to as many observers as possible. An occultation of a bright star will tell us much about the particle density of the rings as well as setting an upper limit to the particle size. If a flickering of the star were observed then we could assume a fairly large particle size; if there were no flickering then we would conclude that the particle size is small.

There are a number of phenomena associated with the rings that the observer can watch out for. One unusual effect is the appearance of radial bands on Ring A. These are light and dark bands radiating from the centre of the planet that seem to cross the rings. This phenomenon is often reported but there is no easy explanation for it. Any flat disc marked with many fine concentric grooves will show radial bands when viewed in oblique light. The question is, 'Are the rings of Saturn marked in this way?' It is doubtful, but the phenomenon is worth looking for. As an analogy, think of a gramophone record displaying radial bands when light focuses on it from a certain angle.

Another unusual phenomenon that has been reported from time to time is what is known as the 'bi-coloured aspect' of the rings. There is occasionally a difference seen in brightness between the ring ansae when they are viewed through a red or a blue filter, although mostly there will appear to be no difference. There has so far been no satisfactory explanation for this and only rarely does there seem any pattern to the results. It has been reported that the effect shows most clearly around quadrature. One particular ansa may then seem persistently brighter through one colour prior to opposition, and the other ansa brighter in the same colour following opposition. But this is far from conclusive and the results are usually very erratic, so that no hard and fast conclusions may be drawn.

Ring plane passage

For the serious student of Saturn the most interesting time of all is the period when the rings are presented edge-on to the Earth. At such times the planet appears stripped of its splendour but the series of phenomena associated with this edgewise phase are so numerous that they almost defy description. This is the time when the normally placid Saturn becomes the scene of great activity.

During the course of one complete revolution of Saturn around the Sun, the plane of the rings will pass through the Sun and the orbit of the Earth on two occasions, spaced about fourteen years apart. The eccentricity of Saturn's orbit causes these passages to be spaced unequally and to occur alternately after thirteen and fifteen years. This is simply the result of the perihelion part of Saturn's orbit being completed more quickly than its aphelion portion.

The conditions for the observation of the ring plane passage may be either favourable or unfavourable as viewed from Earth, but the reasons for this would require a lengthy explanation that is beyond the scope of this book. The ring plane takes just over one year to pass completely through the Earth's orbit and, depending on the position of the Earth in its orbit when the ring plane begins to cross, we will either experience one passage through the plane or three.

If the Earth is on the far side of its orbit, as seen from Saturn, that is, between second and first quadrature when the ring plane first contacts the Earth's orbit, then the first ring plane passage will occur somewhere between conjunction and first quadrature. Following this, the Earth will again catch up with the ring plane, at about opposition, and we will experience a second passage through the plane in the opposite direction. Finally, the ring plane will again catch up with the Earth somewhere between second quadrature and conjunction, and a third passage, in the same direction as the first, will occur. Usually all three passages take place when Saturn is fairly well-placed for observation.

If, on the other hand, the Earth is on the same side of the Sun as Saturn when the ring plane first contacts its orbit, that is between first and second quadrature, then there will only be one passage through the ring plane. In either of the above cases there is, in addition, one passage of the ring plane through the Sun. There are, therefore, four separate events to be seen on favourable occasions and two on unfavourable occasions. With the latter there is an added difficulty. The single plane passage and the passage through the Sun occur close to the time of conjunction and so are rarely observable.

At certain times, the plane will almost pass through the Earth but not quite and, on such occasions, the rings will appear to become edge-on but will then open up again in the same direction. Such an 'appulse' of the rings is rare, but it can happen.

Fortunate period

We are in a very fortunate period for ring plane passages at present. The last occasion of 1966 resulted in three passages, as in fact will the next two occasions of 1979–80 and 1995–96, and so the chances for making observations of this kind of event could not be better. Passages with spring oppositions will find the Earth and Sun passing from the southern to the northern face of the rings, after which there follows a 15-year period with Saturn south of the equator and with the northern face of the rings turned Earthward. Passages with oppositions in autumn take place while the Earth and Sun are passing from the northern to the southern face of the rings, with a following 13-year period when Saturn is north of the celestial equator and the southern face of the rings is directed Earthward. A calendar for the events of 1979–80 and 1995–96 is given below:

1979 October 27	Earth passes south to north.
1980 March 3	Sun south to north.
1980 March 12	Earth passes north to south.
1980 July 23	Earth passes south to north.
1995 May 20	Earth passes north to south.
1995 August 11	Earth passes south to north.
1995 November 19	Sun north to south.
1996 February 12	Earth passes north to south.

In the first case, about seven weeks after conjunction the rings became edgewise and from October 27, 1979, until March 3, 1980, we could view the unilluminated face of the rings. On March 3, the Sun joined us on the northern face and for about a week we observed a very slight brightening of the rings. This terminated with our passage back to the southern face and, from March 12 until July 23, we viewed the unilluminated southern face. Opposition of Saturn occurred on March 14, and so the events in March were very favourable. With the July passage of the Earth through the ring plane we will view the illuminated northern face until 1995.

In the second case, the first edgewise rings occur about six weeks after conjunction, and, from May 20 until August 11, we view the unilluminated southern face. Following the August passage, we see the illuminated northern face once more. On November 19 the Sun leaves the northern face to illuminate the southern face, and so, we will experience a gradual fade of the rings in the first three weeks of November until the rings become completely unilluminated on the northern side. We then view an unilluminated face up until when we join the Sun on the southern face on February 12, 1996. Opposition occurs on September 14, 1995, so the middle two events should be easily observed. However, conjunction which occurs on March 20, 1996, only five weeks after the last event, will make it difficult to observe.

When exactly edge-on the extreme thinness of the rings renders them invisible for a short time. It has been said that with large telescopes it is possible to follow them through a passage, but, unless one views at the very moment of edgewise rings, one is seeing a very foreshortened view of the full ring face. Perhaps a few hours before they become edgewise, the rings are visible in large telescopes, but the author feels that at the exact moment of passage it would tax even the largest telescope in the world to spot them.

Ghostly appearance

The unilluminated face of the rings, seen at the times when the Sun is on the opposite side of the ring plane to the Earth, presents a ghost-like appearance, with two faint patches of light visible on each ansa. These patches, though very faint, can be seen in moderate telescopes and have been estimated at around magnitude +13. The closer condensations seem to be situated in the position of Ring C, while the outer condensations are the Cassini Division area. The inner condensations are apparently the result of sunlight sifting through the rarefied particles of Ring C. The outer condensations are brought about in a similar fashion. The Sun is sifting through the sparsely populated region of the Cassini Division, as Barnard had suggested in 1907. There will be ample opportunity to observe this phenomenon during both the 1979–80 and 1995–96 passages since the Earth spends long periods of time on the unilluminated side of the rings.

Small luminous points are occasionally seen along the thin edgewise ring and there are often considerable differences apparent in the length of each ansa when near edgewise rings. No satisfactory explanation has been provided for either of these phenomena, and both are the subject of present controversy.

Most impressive of all the phenomena is the 'Seeliger effect'. This happens as the Sun passes through the ring plane while the Earth is situated some way north or south of it. For example, if the Earth is south of the ring plane, with the rings appearing slightly open and fully illuminated, the plane of the rings may begin to pass through the Sun. The rings will then gradually fade until they finally disappear, as the sunlight strikes the other face. This fading is a result of the mutual eclipsing of the ring particles. Each particle casts a shadow, which will eclipse other particles in line with both itself and the Sun. As the angle of illumination increases in relation to the Earth, more and more of the eclipsed particles come into view, and this results in a reduction in the brightness of the ring. Something similar is, in fact, seen each year. The rings always appear much brighter at opposition than they do at either quadrature. This is a phase and mutual eclipse effect among the ring particles.

Saturn, like Jupiter, has a considerable family of moons. Ten have so far been discovered and these are generally quite large bodies. Four of them can be seen in fairly small telescopes and up to eight with moderate instruments. They are all rather interesting bodies and observations of them should form a major part of any study programme for Saturn.

The brightest of these moons, and the first to be discovered, by Christiaan Huygens, in 1655 is Titan. Its mean distance from the centre of Saturn is 1,222,000 kilometres and it is possibly the largest moon in the solar system, having a diameter of 5,830 kilometres. Some astronomers believe that Triton, Neptune's larger moon, may well be bigger but this is uncertain. The magnitude of Titan at mean opposition is 8.4, so that it is well within the range of a 50-mm refractor and its sidereal period is 15.9 days.

The next four satellite discoveries were made by J. D. Cassini. Iapetus, found in 1671, has a mean distance from Saturn of 3,560,000 kilometres, a diameter of 1,600 kilometres, a sidereal period of 79.3 days and a mean opposition magnitude of 11.0. Cassini's second discovery, Rhea, was made in the following year. It has a mean distance from Saturn of 527,000 kilometres, a diameter of 1,580 kilometres, a sidereal period of 4.5 days and a magnitude of 9.8. Cassini's next two discoveries came in 1684: Dione, which has a mean distance of 378,000 kilometres from Saturn, a diameter of 820 kilometres, a sidereal period of 2.7 days and a magnitude of 10.4; and Tethys, orbiting at a distance of 295,000 kilometres from Saturn and having a diameter of 1,040 kilometres, a sidereal period of almost 1.9 days and a magnitude of 10.3. These last two satellites are bright enough to be seen with small telescopes but their close proximity to Saturn presents difficulties as they are always immersed in its glare.

More moons

In 1789, W. Herschel made two further discoveries. The first, Enceladus, has a distance of 238,000 kilometres from Saturn, a diameter of 600 kilometres, a sidereal period of nearly 1.4 days and a magnitude of 11.8.

Saturn's triple ring plane passage and the passage of the ring plane through the Sun and Earth.

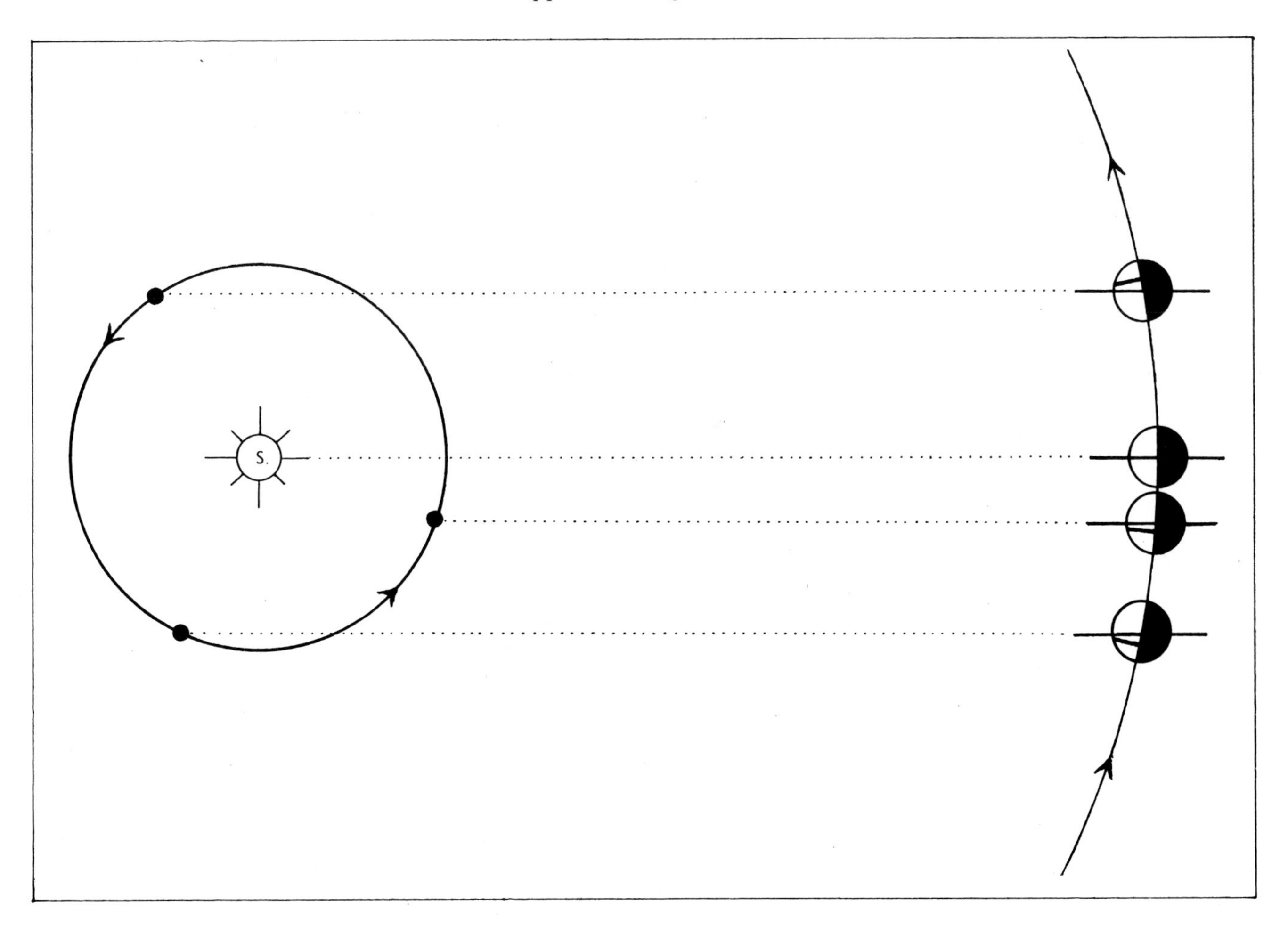

Above left *9.XI.66 20.50 UT 225mm Refl ×274 Alignment of satellites at time of edgewise rings. Note how Iapetus is out of alignment.* Above right *29.I.77 02.30 UT 419mm Refl ×320 The more usual scattered appearance of satellites.*

Mimas, the other, is quite close to the outer ring edge with a distance of only 186,000 kilometres from the centre of Saturn, its diameter a mere 500 kilometres, a sidereal period of only 0.94 of a day and a 12.1 magnitude.

The next discovery did not come until 1848 when the American astronomers, W. Bond and G. Bond, and the English astronomer, W. Lassell, discovered, independently of each other, a satellite with an orbit just outside Titan's. Hyperion, so-called, has a distance of 1,483,000 kilometres from Saturn, a diameter of 500 kilometres (like Mimas), a sidereal period of 21.3 days and a rather faint magnitude at 14.1.

The ninth satellite of Saturn was found in 1898 by W. H. Pickering (known among other things for his observations of the apparent elliptical shape of Jupiter's Galilean satellites), using photographic methods. It is the most distant of Saturn's known satellites and, like the outermost satellites of Jupiter, it has a retrograde orbit, which is also highly inclined and elliptical. Its mean distance from Saturn is an enormous 12,952,000 kilometres, its diameter a mere 200 kilometres, and its sidereal period 550.3 days, or slightly over 18 months. The magnitude is unfortunately only 16.5, and so it is well beyond the reach of most amateur-owned telescopes.

Latest discovery

The last satellite to be discovered was Janus. Though of 14th magnitude, this satellite escaped detection until its discovery in 1966 by A. Dollfus. The reason for this is its close proximity to the brilliant rings of Saturn. Its distance from the centre

of the planet is in fact only 159,000 kilometres, which is about 23,000 kilometres from the outer edge of Ring A. The satellite was found during the edgewise ring presentation of 1966, when the glare from the rings was non-existent. Even then an occulting bar had to be used to reduce the glare from the disc of Saturn itself. Janus' orbital period is a very rapid 17 h 55 m, but virtually nothing more is known about the satellite and it has not been seen since 1966. The edgewise presentation of the rings in 1980 should enable us to obtain further information. Examination of the photographs showing Janus would suggest that there might be one or more satellites orbiting even closer to the rings than Janus.

Amateur observation of the satellites is at present concerned mainly with the determination of their magnitudes, since they all seem to show some variation. Most interesting in this respect is Iapetus. It always appears brighter at western elongation than at eastern elongation and the difference amounts to more than two magnitudes. It is evident that one hemisphere of the satellite reflects much more efficiently than the other —perhaps, for some reason, the satellite is coated with ice on one side and not the other. The regular nature of the light variations also indicates that the satellite has a captured rotation. The amplitude of the variations, however, does alter. No doubt this is in some way due to the changing tilt of the satellite, which presumably changes with the tilt of Saturn.

Determining magnitudes

Rhea, too, displays marked variation in brightness and usually appears brightest at eastern elongation. All the remaining brighter satellites seem to show some variation and there is scope here for the amateur to make a contribution. The problem is that there are no fixed reference stars for the determination of the satellites' magnitudes. All that can be done is

26.VIII.66 23.20 UT 225mm Refl ×274 Rings of Saturn with the shadow of Titan visible.

to compare their magnitudes with each other. This is not particularly satisfactory because they are all, of course, constantly changing and their varying distances from Saturn make matters more difficult still. Nevertheless, this type of observation poses an interesting challenge to even the most assiduous observer.

Titan's colour

Studying the colour of Titan is a further valuable exercise. Generally, this satellite has a distinct orange colour, but there have been reports of variation in the colour. Observations with colour filters also show variations in its visibility. Sometimes it will appear brighter when viewed in red light, sometimes in blue light. On the whole, the filter results seem to show it, brighter in blue light more often than in red. This is the opposite of what one would expect from the colour of the satellite itself. No connection between these variations and the satellites orbital period has been established. Titan, however, seems to have a considerable atmosphere and the explanation is perhaps, to be found there.

Titan is the only satellite of Saturn that shows an appreciable disc. At just under 1″ it is easily resolvable with moderate telescopes. Observers using large telescopes have detected markings on the satellite but, in general, these markings do not seem to be of a permanent nature. This suggests that they are atmospheric. The author has often seen the satellite resolved clearly into a disc with magnifications of about ×400 and has repeatedly been impressed by the considerable limb darkening, which is further suggestive of an atmosphere.

However fascinating the satellites themselves may be their phenomena in relation to Saturn are even more so. The period when the plane of the rings passes through the Earth's orbit, and through the Sun, is accompanied by a passage through the plane of the orbits of most of the satellites. Indeed, for a couple of years either side of edgewise rings, we see the phenomena of satellite occultations, eclipses, transits and shadow transits, all of which are commonly seen on Jupiter. The circumstances for these events is actually the same as described for Jupiter but eclipse timings are particularly important because of their rarity.

Spectacular phenomena

Titan, the largest of Saturn's satellites, lays claim to the most striking phenomena. Titan's shadow is especially clear and has a prominence close to that of the shadow of Io on Jupiter. It can easily be detected with only a 75-mm refractor. Transits of Titan are also of interest since, when superimposed on the bright disc of Saturn, the satellite shows up as a small brownish grey spot. Phenomena of the other satellites are of course less spectacular. Rhea does not present too many problems; its shadow has been glimpsed with a 150-mm reflector. Occultations and eclipses of Tethys and Dione are also quite easy

29.X.66 20.05/21.35 UT 225mm Refl ×274 Above *Titan shows as a dusky spot.* Below *90 minutes later it has left the disc while its shadow has moved on to the disc near the North Polar Region.*

with moderate-sized telescopes. Transits and shadow transits for these and the other inner satellites are, however, very difficult to observe. Investigations into the limiting apertures required for the observation of events which involve the smaller satellites are being conducted by most bodies concerning themselves with Saturn.

Undoubtedly one of the most useful and interesting observations of this type concerns the satellite Iapetus. The orbit of Iapetus is not in the plane of Saturn's equator but inclined at nearly 15° to it. This means that any phenomena for this satellite do not occur at the same time as for the others but usually about two years before the edgewise presentation of the rings, that is, when the rings are fairly wide open. Occasionally the satellite may pass through the shadow of the rings, which gives us an opportunity to see just how much light passes through them. One of the most famous observations of this type was carried out by E. E. Barnard in 1889. Using a 300-mm refractor, he observed an eclipse of Iapetus by Ring C and was able to watch the satellite throughout the eclipse. Unfortunately, because of the duration of the event, he was unable to observe it in full and did not see the satellite eclipsed by Saturn's other rings.

Eclipse of Iapetus

The last opportunity to make a comparable observation occurred during the 1977–78 apparition. There were, in fact, two chances since eclipses took place both on October 19, 1977, and January 7, 1978.

On account of Iapetus' slow rate of motion, each full event took around 16 hours. The satellite was never lost while in the shadow of Ring A and a slight brightening at that time confirmed the existence of Enckes Division—the only other generally accepted division in the rings besides Cassini's. As has been said before these events are regrettably very rare.

There is a lot of speculation and controversy on the subject of the rings of which the amateur should be aware. It has been suggested, for example, that the rings do not end suddenly at the visible outer edge of Ring A but fade off gradually into space. Reports of an exterior ring (Ring E) have been

29.III.78 20.20 UT 419mm Refl ×372 Ring detail, suggesting varying particle density.

made from time to time but it is by no means a generally accepted feature. There have also been odd reports of a faint ring (Ring D) inside Ring C, but again this is controversial. However, the possibility of the existence of ring particles inside Ring C was a contributory factor to the decision taken by the Americans not to allow their Pioneer 11 spacecraft to pass between the rings and the globe during its 1979 close encounter. The Equatorial Band of Saturn has been put down to the possible shadow of such an inner ring, and this is currently under investigation but has not yet been proved.

Subtle shades

The colours of Saturn, though less exciting than Jupiter, have been the subject of a number of reports and certainly deserve a mention.

The belts are usually brownish grey to grey and occasional red-brown tints can be seen. The zones are yellowish to cream or sometimes pale buff or fawn. The polar regions may from time to time assume a greenish or olive coloration, which is perhaps nothing more than combined effect of the yellows and greys. The Equatorial Zone seems to vary in colour to some extent and can be either white, yellow or even pale pink. The rings themselves do not exhibit too much colour. Ring B tends to be silvery white or pale cream; Ring A silvery grey, with perhaps a yellowish tint; and Ring C, though basically grey, seems to oscillate between brownish grey and bluish grey. The overall picture is one of great beauty even though the colours and various shades are very subtle.

Saturn is, at present, the most distant of the planets to have been visited by a space probe. Pioneer 11 passed within 21,000 kilometres on September 1 1979, after a 6½-year journey. Among its early discoveries were three new satellites and at least two more rings beyond the main rings. Evidence of jet streams in the planet's atmosphere have also been found. In passing Saturn, Pioneer has blazed a trail for the two Voyager probes due to arrive in 1980 and 1981.

Voyager 1 is expected to make an approach to within 130,000 kilometres of the planet, and, if all goes well, 4,000 kilometres of Titan. Close approaches to the satellites Mimas, Dione and Rhea are also expected. Finally, Voyager 2 will make its close approach on August 27, 1981, to pass within 40,000 kilometres of the ring edge and have close encounters with Titan, Rhea, Tethys, Enceladus and Mimas. As astronomers, we have an exciting time ahead of us and, if the picture of Jupiter and its moons returned by Voyager are any indication of what can be achieved, there should be many startling revelations about Saturn.

14.IV.77 22.00 UT 419mm Refl ×372 Titan's fuzzy disc suggests a considerable atmosphere.

Even without the help of spacecraft we have been able to build up quite a good idea of the conditions that exist on, and within, Saturn. We know that the planet, like Jupiter, radiates far more heat than it receives since temperature readings of the cloud tops give between −145°C and −180°C, which is similar to Jupiter despite Saturn's greater distance from the Sun. This heat from within Saturn probably drives the convection in the atmosphere.

Astronomers believe that at the centre of Saturn there is a rocky core measuring 20,000 km in diameter, which is much larger than Jupiter's. Though hot, this core is surrounded by a 5,000-kilometre thick envelope of ice which is, in its turn, enclosed in an 8,000-kilometre shell of metallic hydrogen. All of this is contained in a deep envelope of liquid hydrogen, which forms the greater bulk of Saturn. Outside this, there is a shallow, fairly dense atmosphere of hydrogen, helium, methane and ammonia. The ammonia content is, however, far less than in Jupiter's atmosphere.

Fine spectacle

Studies carried out on the rings of Saturn, using both infra-red and radar methods suggest that they are composed of water and ammonia ice crystals together with dust and rock fragments. Almost the whole of the inner half of the rings revolve around Saturn in a shorter time than the planet itself takes to rotate, but the outer half revolves more slowly. Particles at the inner edge of Ring C have a period of less than 5.5 hours, while particles at the outer edge of Ring A have a period of around 14 hours.

If the motions of the ring particles could be observed from the planet they would present a fine spectacle. The inner edge would appear to move slowly eastward while the outer edge would move slowly westward. Viewed from the vicinity of the planet's equator, the rings would appear as a brilliant arc in the sky. On the night sky, this arc would be very often broken by the huge shadow of the planet but it would eclipse the stars and form a black band across part of the sky.

When the rings were wide open with the Sun shining on them over the pole of the planet, they would appear unbroken and would form a magnificent bright arc across the sky.

Most of Saturn's inner satellites are believed to be composed largely of an ice and rock mixture. Infra-red observations have shown that at least the four moons Tethys, Dione, Rhea and Iapetus, all have surfaces covered with water ice. The covering on Iapetus tends to be patchy, which could account for the satellite's light variations.

Appreciable atmosphere

The oddest of these worlds must surely be Titan. It, too, has a rock and ice composition, but it is believed to be the one satellite in the Solar System that has an appreciable atmosphere—thought to have a surface pressure of between one and one-tenth of the Earth's atmospheric pressure at sea level. The composition of the atmosphere is methane and hydrogen, which is not unlike that of its primary. The orange clouds, that give Titan its colour may well result from a photochemical reaction of particles suspended in this atmosphere. But how Titan retains its atmosphere is puzzling. It seems that there must be a continual 'outgassing' from within the satellite, which keeps it replenished, or possibly even a certain amount of liquid methane on the surface of the satellite.

The interior of Titan is equally interesting if what astronomers believe is correct. A small central rocky core is thought to be surrounded by a fairly thick envelope of water ice and rock. This is encased in a thin rocky crust outside of which there is a thin methane and ice coating. Titan is a larger body than Mercury and it is a pity that it lies at such a great distance from us. It will certainly be a major object for study by the space probes visiting the planet.

Saturn is a glorious spectacle and with its cloud belts, spots, rings and variable satellites it offers the would-be observer a greater variety of phenomena than any other body in the Solar System.

Imaginary view of Saturn from the satellite Iapetus, with the ring broken by shadow.

Uranus

On the night of December 23, 1690, John Flamsteed, the first Astronomer Royal, was charting the stars in the constellation of Taurus. One of the stars he charted was of the 6th magnitude and he designated it 34 Tauri. Had he taken the trouble to check this star's position on subsequent nights he might have made a discovery that would have shocked the astronomical world. Flamsteed's 34 Tauri was not a star but a planet and it would have been seen to move from night to night against the background of the stars.

Twenty-two recorded observations of this object were made between 1690 and 1781. Flamsteed himself recorded it a further five times in different parts of the sky, each time thinking it a different star. The remaining observations were made by three other notable astronomers of the period. Not one of them realized the nature of the object they had seen. Then, on the night of March 13, 1781, a brilliant, but at the time amateur, astronomer William Herschel was charting stars in the region of the Gemini–Taurus border between Messier 1 (Crab Nebula), and the beautiful open cluster, Messier 35. He was taken by the appearance of a 6th magnitude star close to what is now known as 132 Tau. This object did not have the tiny point image of a star but appeared as a small disc. The fact that he immediately noticed this with a telescope of only 150 mm was a credit to his skill as an observer and to the quality of the instrument used, which was one that he had constructed himself.

Startling discovery

Further observations of the object soon showed that it was moving slowly among the stars and at first it was thought that it might be a comet. Eventually it was found that its motion did not resemble the motion of a comet at all and the startling realization slowly came to astronomers that this was a new planet, the first to be discovered since the dawn of history. One can only try to imagine the effect that this discovery had on astronomy at the time.

Imaginary view of Uranus from its moon Miranda. Far from the Sun, it is a very cold world.

After a while, the planet was given the name Uranus. Its symbol, ♅, commemorates its discovery by Herschel.

Uranus was found to be a very remote planet. Its distance of 2,869.6 million kilometres doubled the size of the known Solar System. Although Uranus is a large gas planet it is much smaller than either Jupiter or Saturn and present estimates put its diameter in the region of 52,000 kilometres. There is, however, a great degree of uncertainty about this, which is hardly surprising considering its distance from us. Some estimates put its diameter at only 47,000 kilometres. Roughly it is about four times the diameter of the Earth. Its volume is around 60 times the Earth's, but its mass only 14.6 times the Earth's mass. So, like Jupiter and Saturn, its density is very low, about 1.2 times the density of water.

Due to its great distance from the Sun, Uranus has a long orbital period of 84 years and one week. Strangest of all is its axial tilt, which is 98°. In fact, Uranus is lying on its side. Its rotation, although actually direct or anti-clockwise, as viewed from a point above its north pole, appears retrograde to us because it is tipped past the horizontal. If its axial tilt were considered as 82° then the rotation would indeed be retrograde. All this has a peculiar effect on Uranus' 21-year seasons. Sometimes the equatorial regions are directed towards the Sun, at other times either one or other of the poles will be directed towards the Sun. This means that each pole will suffer 21-year long nights alternately.

Rotation period

Like most other things about this planet, its rate of spin is not known accurately. Until recently it was generally accepted that its day was 10 h 49 m long. New observations have suggested a much longer period of more than 22 hours. Whatever its period, Uranus shows a marked polar flattening, resulting from a fairly rapid spin and its low density. Estimates put its polar diameter at 3,000 kilometres less than the equatorial diameter.

With a magnitude averaging +5.7 Uranus is just visible to the naked eye. A fair amount of magnification is needed to give the 4″ diameter disc a sizeable image. A 75-mm refractor with a power of ×60 should just about show it, but a magnification of ×450 and a moderate aperture telescope are required to give it a clear image about the size of the Moon to the unaided eye. The disc itself does not show much in the way of detail.

Occasionally, faint belts and a bright equatorial zone have been detected which resemble generally the appearance of Saturn, but with nowhere near the clarity. There seems to be very little minor detail in the belts. On the other hand many observers, using quite large telescopes, report a complete absence of any belt structure and find the planet displays irregular dark and light patches. Because of the changing orientation of the planet as seen from Earth, any belt structure, if it exists, could only

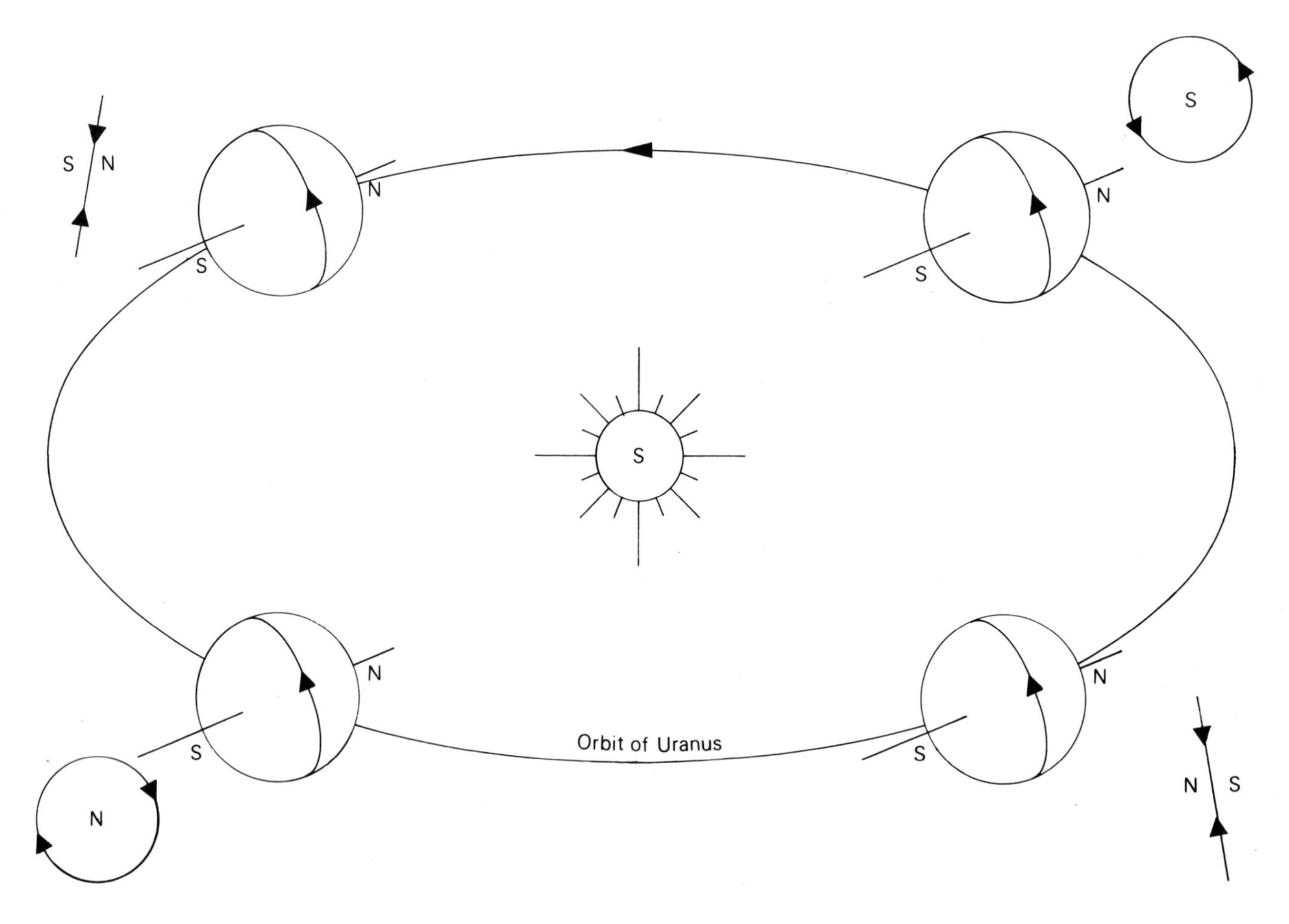

Why Uranus appears pole-on at certain times while at others its equator is presented Earthward.

be expected when the planet's equatorial region is directed towards us. The last time this happened was in 1965. When either pole is directed towards us no belts should be seen. This generally seems to be the case. Observations made when the pole is directed towards us show a dark spot in the centre of the disc, suggesting a darkened polar region. At present, Uranus' northern pole is turning towards us and will be central on the disc in 1985. This means that we may not expect to see any belts until the 1990s. The equatorial plane of Uranus will not pass through the Earth's orbit again until the year 2004.

Mostly blue

The colour of Uranus, when viewed through a telescope, is distinctly bluish green. The atmosphere of the planet is composed largely of hydrogen, helium and methane. Most of the ammonia seems to have been frozen out and consequently there are no opaque ammonia clouds. This means that sunlight can penetrate deep into the atmosphere and, since methane strongly absorbs light of longer wave length—red and yellow—the light reflected back by Uranus is mostly blue, hence the visual colour. The absence of ammonia clouds probably results in the lack of detail displayed by the planet. Even high resolution photographs taken with telescopes that have been lifted high in our own atmosphere by balloons show a virtually blank disc.

The flattening of the disc is visible with only moderate magnification, but again only when the equator of the planet is directed towards us. When a pole is facing us the disc is perfectly round. One marked feature is the considerable limb darkening, which is to be expected with a planet that has such a considerable atmosphere. This is so evident that it is difficult to get a good focus on the planet alone and really this should not be attempted. It is far better to focus on a nearby star before observing the planet.

There is one type of observation that the amateur can make and that is the estimation of the planet's magnitude. A large telescope is not needed for this sort of work and, in fact, small telescopes or binoculars are far better since they give a point image.

6.VI.78 23.20 UT 385mm Refl ×350 Only faint light and dark patches may be seen on Uranus.

It is quite easy to compare the magnitude with the surrounding stars and over a period of time fluctuations will show up. Estimates of the planet's rotation period have been made in this way but there have also been irregular, long, period fluctuations reported. It is possible that these may result from changes within the planet's atmosphere. Obviously, there will be changes due to the planet's varying distance from us. Most interesting is the fact that brightness variations in this planet are being used as an indication to brightness changes in the Sun. Any small alteration in the brightness of the Sun would be difficult to detect simply because it is so bright but if there are any changes these would be reflected in the brightness of the planets, and the remote Uranus seems particularly suited. An experienced amateur could do useful work in this field.

Even if an observer does not particularly wish to become involved in any of the work mentioned he could keep himself happy by simply charting the planet's motion among the stars. Again, only a small telescope is required and it is quite pleasing to watch the planet's nightly movement. There is always the chance that the planet may occult a star and this could be of real importance.

Family of satellites

Uranus has a fairly large family of satellites. Five have so far been discovered and these form a very regular system. The outer two moons were found by Herschel himself during his observations of 1787. The outermost, Oberon, has a distance from Uranus of 586,000 kilometres. Its orbital period is slightly over 13 days, 11 h 15 m, its diameter in the order of 1,500 kilometres and its mean opposition magnitude 14.2. Titania is 438,000 kilometres distant, has a period of 8 days 17 hours, a diameter of 1,800 kilometres and a magnitude of 14.0. The inner two of the brighter satellites were discovered by William Lassell. Their discovery was announced in 1851 but he suspected the existence of two inner satellites three years previously. The outermost of these inner two is Umbriel, which has a distance from Uranus of 267,000 kilometres, a

2.VI.78 23.30 UT and 6.VI.78 23.15 UT 385mm Refl ×350 The satellites of Uranus are not easily seen, because they are faint and close to the planet.

period of 4 days 3 h 28 m, a diameter of between 800 and 1,000 kilometres and a magnitude of 15.3. Ariel has a distance of 192,000 kilometres, a period of 2 days 12 h 29 m, a diameter of between 1,000 and 1,500 kilometres and a magnitude of 14.4.

The fifth satellite of the planet was discovered quite recently. G. P. Kuiper, while taking photographs of Uranus' brighter satellites with the 2-metre McDonald reflector, in 1948 detected another faint inner satellite. Subsequently named Miranda, this has a distance of only 131,000 kilometres from Uranus, and an orbital period of only 1 day 9 h 55 m. It has a small 300–500-kilometre diameter and a magnitude of little more than 17.

The outer two satellites are within the reach of moderate amateur equipment. Their discovery was made with Herschel's 450-mm reflector but it has been claimed that they are visible with a 150-mm refractor when Uranus is well placed. To observe them properly requires at least a 300-mm reflector and with the planet poorly placed from Britain, the author finds them difficult with a 419-mm reflector. The two inner satellites are much more difficult and these were discovered with a 600-mm reflector. The inner of the two is of similar brightness to Oberon and Titania but its closeness to the planet make it very difficult. The outer of the two is slightly easier to observe but, of course, it is fainter than the rest. The author has just about glimpsed it with a 419-mm reflector. Finally, Miranda is well beyond the reach of the equipment of an amateur observer.

Uranus' rings

Observations of these satellites have proved them to be slightly variable. It is not known if their equators lie in the same plane as their parent planet's. If they do, then we would see them alternately pole on—equator on, in the same period as Uranus. This could result in their brightness variation but, because of the difficulties involved this is far from conclusively proven.

The motions of the satellites, as seen from the Earth, are of interest. Since their orbits lie in the plane of the planet's equator they sometimes appear to move to and fro in a straight line, but when the poles face us their orbits describe a full circle. When the planet's north pole is facing us, as at the present time, they appear to revolve round the planet in an anticlockwise direction, but when the south pole is presented to us their motions appear clockwise. Their present anticlockwise motion will continue until the passage of their orbit plane through the Earth's orbit in 2004.

Perhaps the most fascinating discovery of recent times is the fact that Uranus has a system of rings. These

10 and 11.III.77 01.00 UT 419mm Refl ×176 The night prior (above) and after (below) the occultation of the star SA0158687 by Uranus, which led to the discovery of its rings.

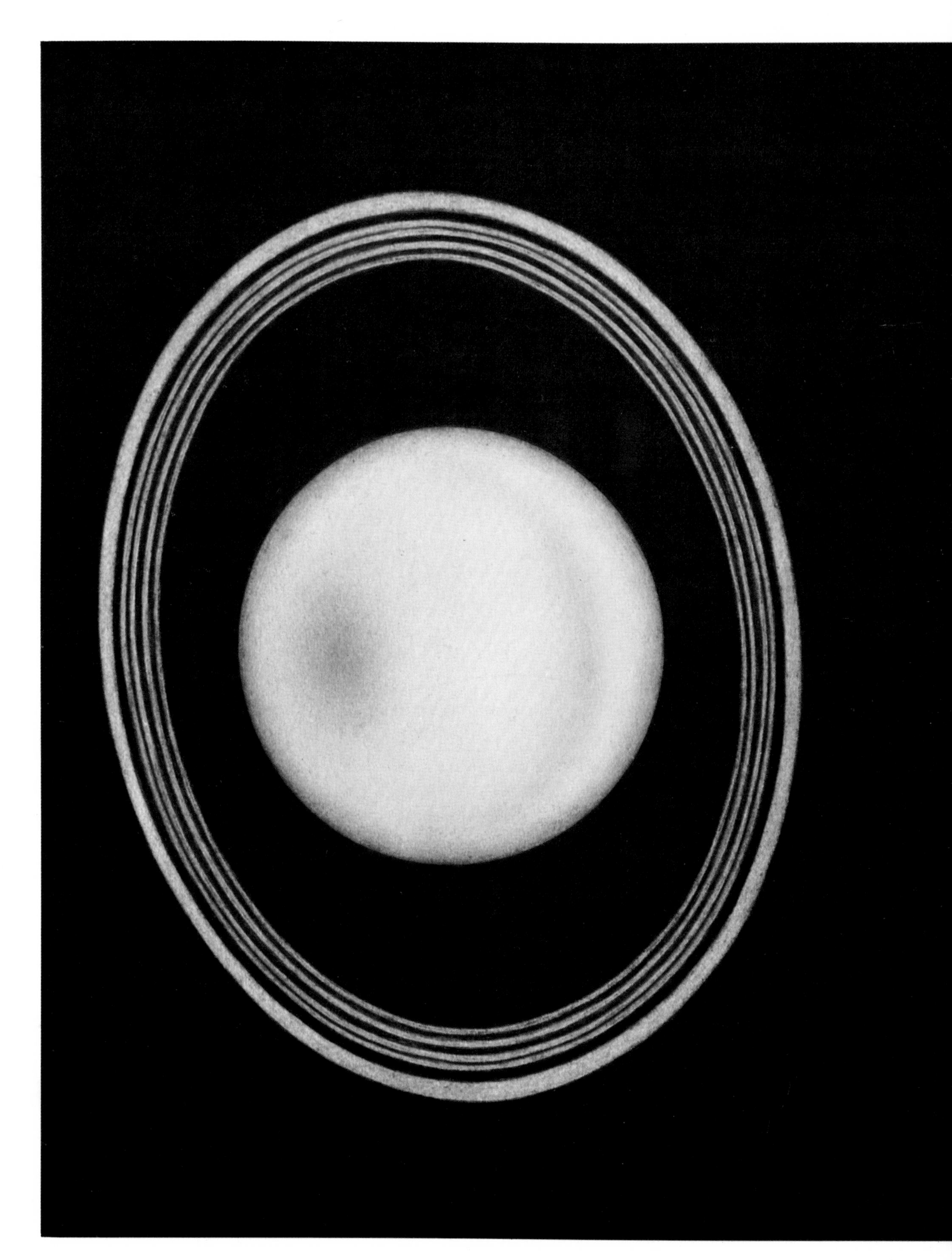

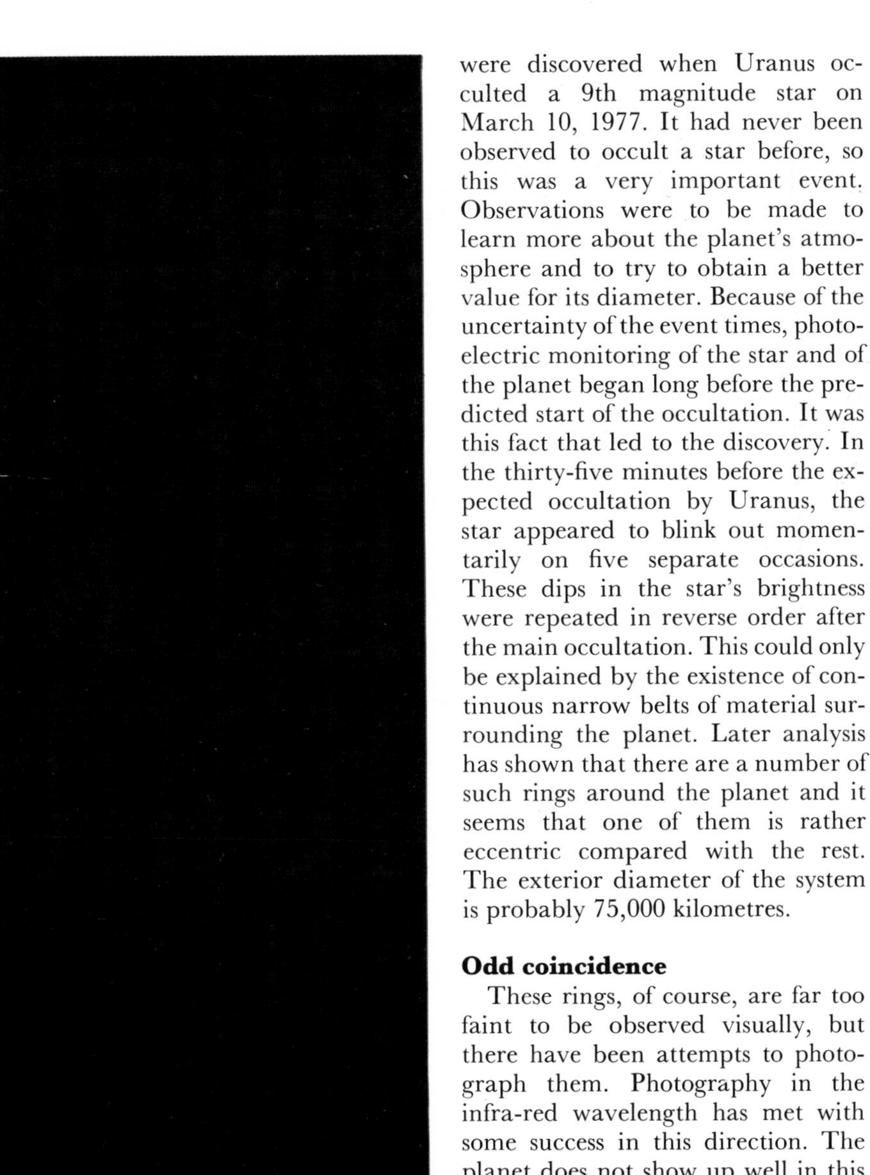

were discovered when Uranus occulted a 9th magnitude star on March 10, 1977. It had never been observed to occult a star before, so this was a very important event. Observations were to be made to learn more about the planet's atmosphere and to try to obtain a better value for its diameter. Because of the uncertainty of the event times, photoelectric monitoring of the star and of the planet began long before the predicted start of the occultation. It was this fact that led to the discovery. In the thirty-five minutes before the expected occultation by Uranus, the star appeared to blink out momentarily on five separate occasions. These dips in the star's brightness were repeated in reverse order after the main occultation. This could only be explained by the existence of continuous narrow belts of material surrounding the planet. Later analysis has shown that there are a number of such rings around the planet and it seems that one of them is rather eccentric compared with the rest. The exterior diameter of the system is probably 75,000 kilometres.

Odd coincidence

These rings, of course, are far too faint to be observed visually, but there have been attempts to photograph them. Photography in the infra-red wavelength has met with some success in this direction. The planet does not show up well in this light but the satellites and ring material do. Signs of the ring crossing the globe also show on pictures taken with a 1-metre telescope lifted high into the Earth's atmosphere by the Stratoscope II Balloon in March 1970. Generally, the material of the rings seems to be very dark and possibly consists of rock fragments or dust of very low reflectivity. No water or ammonia ice has been detected, so they are quite different from the rings of Saturn. It also seems that they take their present form because of the perturbations of Uranus' satellites.

An imaginary view of the rings of Uranus. The actual rings do exist but are not visible from Earth. This is how they would probably appear if they were bright enough to see. The American Voyager spacecraft will be approaching the planet in 1986 and may send back similar pictures.

It is strange to recount, in the light of this discovery, that shortly after William Herschel discovered Uranus he observed what he thought was a ring round the planet. His observations of this ring were made between 1787 and 1794. Herschel himself eventually stated that his observations were erroneous and that a ring did not exist. It would be very unwise to suggest that Herschel's ring was the predecessor of that which has been recently discovered, but it is certainly a very curious coincidence.

Close encounter

Uranus is a very odd sort of place. No spacecraft has ever ventured there although one may already be on its way. The Voyager 2 spacecraft should carry on to Uranus following its 1981 encounter with Saturn, but this will not be decided finally until the success of the Saturn–Titan encounter of Voyager 1 in 1980. If the Uranus trip does take place, the close encounter should occur in January 1986. There will then be a wonderful opportunity to see the movements of the atmosphere and, possibly, a view of the rings will be obtained.

Until Voyager, our knowledge of Uranus will remain fairly limited. Nevertheless, considering the problems, we are building up a fair idea of conditions on, and inside, Uranus. The planet is thought to have a 16,000-kilometre diameter core of rock, not unlike Saturn's in size. This is enveloped in an 8,000-kilometre deep ice coating which is in turn enveloped in liquid hydrogen. Outside this is, of course, the planet's fairly deep atmosphere. Temperatures at the surface, or rather in its upper atmosphere, are thought to be about −200°C, which is again not too low considering the enormous distance of the planet from the Sun. At such temperatures methane would remain in the form of a gas but ammonia would be frozen.

Because of the difficulty of observation most amateurs leave Uranus alone. With its faint markings, large family of moons and the added importance of any possible occultations of stars by Uranus, perhaps this planet is worth a little more than just a casual glance.

Neptune

Following the discovery of Uranus, it did not take astronomers long to check back through past records and find the observations of the planet made before its discovery, which were mentioned at the beginning of the last chapter. Normally these would have given astronomers of the day everything they needed to enable accurate orbital elements to be worked out. In 1821, a mathematician named A. Bouvard attempted this and discovered that he could not reconcile observations made since the planet's discovery 40 years earlier, with those made in the 90-year period before its discovery. Believing that the earlier observations were probably inaccurate, Bouvard decided to base his calculations on just those observations made since the discovery of Uranus. Yet problems still remained and Uranus refused to comply with its computed motion.

By 1845, two mathematicians, J. C. Adams in England and U. J. J. Leverrier in France, had independently considered the problem and had reached the same conclusion: that some planet, exterior to Uranus, must be exerting its pull and causing the latter to fall behind its expected position. Adams had calculated the position of such a planet by 1845, but his work was not properly followed up. Leverrier completed his calculations in 1846. These were investigated by the astronomers J. G. Galle and H. D'Arrest at the Berlin Observatory. The result was the discovery of Neptune, on December 23, 1846, the very first night of their search. The planet was actually found only one degree from Leverrier's calculated position, which is truly remarkable.

Neptune's distance from the Sun is an unimaginable 4,497 million kilometres, which is half as far again as Uranus is from the Sun. Neptune's orbital velocity is slow and its year is nearly 165 of our years. The planet itself is very similar to Uranus in that it is another gas planet. Its diameter is now estimated at around 49,000 kilometres. Until recently it was thought that Neptune was larger than Uranus, but recent measurement has indicated that this is not so. Its volume is about 54 times that of the Earth and its mass just over 17 times the Earth's. Although, in fact, a little smaller than Uranus it is slightly more massive but its density, though greater than Uranus' is still low, at only 1.8 times that of water.

The axial tilt of Neptune is almost 29°, which is similar to that of Saturn. Its rotation period is at present under review. Until recently, the accepted period was 15 h 48 m, but new spectroscopic observations have indicated that, like Uranus, the rotation period is longer, more like 23 hours with a possible error of 4 hours either way. The flattening of Neptune's poles is not as marked as it is for the other three gas planets on account of its greater density. The polar diameter is calculated at only some 1,000 kilometres less than its equatorial diameter.

The magnitude of Neptune is +7.8, which puts it well below naked-eye visibility. It can be seen with binoculars, but to see it clearly a small telescope is needed. A 150-mm refractor with a power of at least ×150 is really necessary to resolve its 2.3″ disc. To give an image comparable with the Moon requires both a power of ×700 and obviously a telescope of considerable aperture, so there is little chance of amateurs obtaining a good view.

Triton and Nereid

The appearance of the planet even when viewed with a good telescope is only that of a pale blue and featureless disc. Virtually nothing in the way of detail can be seen. The constitution of its atmosphere is similar to that of Uranus. Virtually all the ammonia must be frozen out and even some of the methane is in liquid form. Nevertheless, there is still a considerable amount of methane gas, which absorbs red and yellow light, and gives rise to the planet's blue colour.

Professional astronomers with large telescopes have recorded mottling and faint dusky patches that are irregular in shape. Only occasionally has any belt structure been reported and this must be regarded with suspicion, but a brightish equatorial region has frequently been seen with very large telescopes. The absence of detail is no doubt due to the virtual lack of cloud in the atmosphere. There is a considerable limb-darkening, which tends to give the planet an out-of-focus appearance. One type of observation the amateur can make is to determine the planet's magnitude. Neptune, like Uranus, shows slight brightness fluctuations. Other than this the amateur can just watch the planet's slow nightly movement among the stars.

Two satellites have been found in orbit around Neptune. Triton, the larger and closer, was discovered by William Lassell less than one month after discovery of the planet itself. Its distance from Neptune is 355,000 kilometres and its orbital period 5 days 21 hours. Triton is a large body, having a diameter which is estimated at between 4,000 and 6,000 kilometres. If the latter value is correct then it is the largest satellite in the Solar System, but the size is not known with sufficient accuracy to be sure. It must be very similar in size to Saturn's largest satellite, Titan. Its magnitude of +13.5 is quite bright considering its enormous distance.

In 1949, G. P. Kuiper discovered the second satellite, Nereid, by photography, using the 2-metre McDonald reflector. Nereid is a small body with a very eccentric orbit. Its distance from Neptune ranges between 1.6 million to 9.6 million kilometres. Its period is just under 360 days and it has a diameter about 300 kilometres and has a magnitude of only +18.7.

While Nereid is beyond the reach of amateur instruments, Triton is much easier to see than any of Uranus' satellites. First of all, because it is brighter and secondly because the glare from Neptune is far less than that from Uranus. This satellite should be just about visible with a 250-mm reflector. The author found it easy with a 419-mm reflector, even when Neptune is situated close to the horizon.

It is the orbit of Triton that is odd. Its motion around Neptune is retrograde and it is tilted 35° to the planet's orbit plane. This orbit is possibly changing and Triton may be getting closer to Neptune. One day it may approach so close that tidal forces will cause it to break up and spread around the planet, forming a ring. In fact, Neptune could already have a ring, like Uranus, though there is as yet no observational evidence to suggest this. Obviously, since the discovery of Uranus' rings, which followed upon the planet's occultation of a star, an occultation by Neptune is eagerly awaited. There may be one of two chances of this during the present century but most of the stars concerned are rather faint. The best is an occultation of a 12th magnitude star in February 1980.

By far the best opportunity for such a discovery should have been the occultation of an 8th magnitude star that occurred in 1968. A good value for the diameter of the planet was obtained from it but the possible existence of a ring was not considered and so no evidence for it was sought.

There are reports that two astronomers, William Lassell (discoverer of Triton) and George Challis (unsuccessful searcher for the planet), detected a ring visually during the few weeks following the discovery of Neptune, but this has never been confirmed and cannot be taken seriously.

Cold and distant

Because Neptune is so far away, our knowledge of it is bound to be sketchy. We have built up a kind of picture of its structure and we assume that it resembles Uranus. A rocky core, 16,000 km in diameter, is thought to be surrounded by an 8,000-km thick ice envelope which is, in turn, contained by a deep layer of hydrogen, helium, methane and ammonia. Temperatures in the upper atmosphere are around —220°. Recent observations in the infra-red wavelength have indicated that there are changes in the state of the atmosphere, which probably take the form of some short-term high-altitude cloud formation.

To try to determine just what is going on with a planet so distant is, of course, a very difficult occupation. Currently it is hoped that Voyager 2 might be able to travel on to Neptune after its proposed encounter with Uranus in 1986. There are obviously many maybe's in this, but it cannot be denied that such a visit would prove of interest. If it does materialize the expected arrival time for Voyager would be in the late 1980s. At present, this seems to be the only way that we will obtain any detailed information about this cold distant planet.

6.VI.78 23.48 UT and 9.VI.78 23.45 UT 385mm Refl ×350 Neptune and Triton. Markings on this planet are virtually impossible to see. Triton is easier to spot than the satellites of Uranus, being slightly brighter.

An imaginary view of Neptune from Triton. Somewhat bleak and uninviting, it is very similar to the view of Uranus in many ways.

Pluto

In order of mean distance from the Sun, Pluto is the furthest away of the planets but, as from January 21, 1979, it was no longer technically correct to say that Pluto is the outermost planet. For on that date Pluto crossed the orbit of Neptune and in so doing temporarily forfeited this status. It will remain on the inside of Neptune's orbit until March 1999. The reason for this is that Pluto has a very eccentric orbit and although its mean distance is 5,900 million kilometres from the Sun, its aphelion distance is 7,375 million kilometres and its perihelion distance is 4,425 million kilometres.

We are fortunate to be living in a time when Pluto is approaching its perihelion. This only happens once every 248 years which is the time Pluto takes to complete a circuit of the Sun. A quick glance at the above distances will show that Pluto is nearly twice as far away at aphelion than at perihelion, so our view of it at present is as good as it can ever be.

Tiny disc

The magnitude of the planet is now 13.8 and it should be visible with a 250-mm reflector. When at aphelion it will be much fainter at +16 and therefore beyond the reach of the average amateur-owned telescope. The angular diameter of Pluto is slightly over 0.2″ and only the largest telescopes in the world will resolve its tiny yellowish disc. Because it is so small it is difficult to measure accurately and so the actual diameter of the planet is uncertain. Until recently it was thought to be around 6,000 kilometres or between the sizes of Mercury and Mars. However, in June 1978, the astronomer J. W. Christy noticed while examining photographs of the planet that the image at times appeared slightly elongated. It was found that Pluto has another smaller body in orbit around it, having a revolution period of 6.4 days. Light variations of Pluto itself show this same period which indicates that its rotation period is also 6.4 days. The similarity of the two periods is no doubt a result of tidal friction.

Imaginary view of the Sun as seen from Pluto. The Sun is so distant from the planet that it appears as nothing more than a very brilliant star.

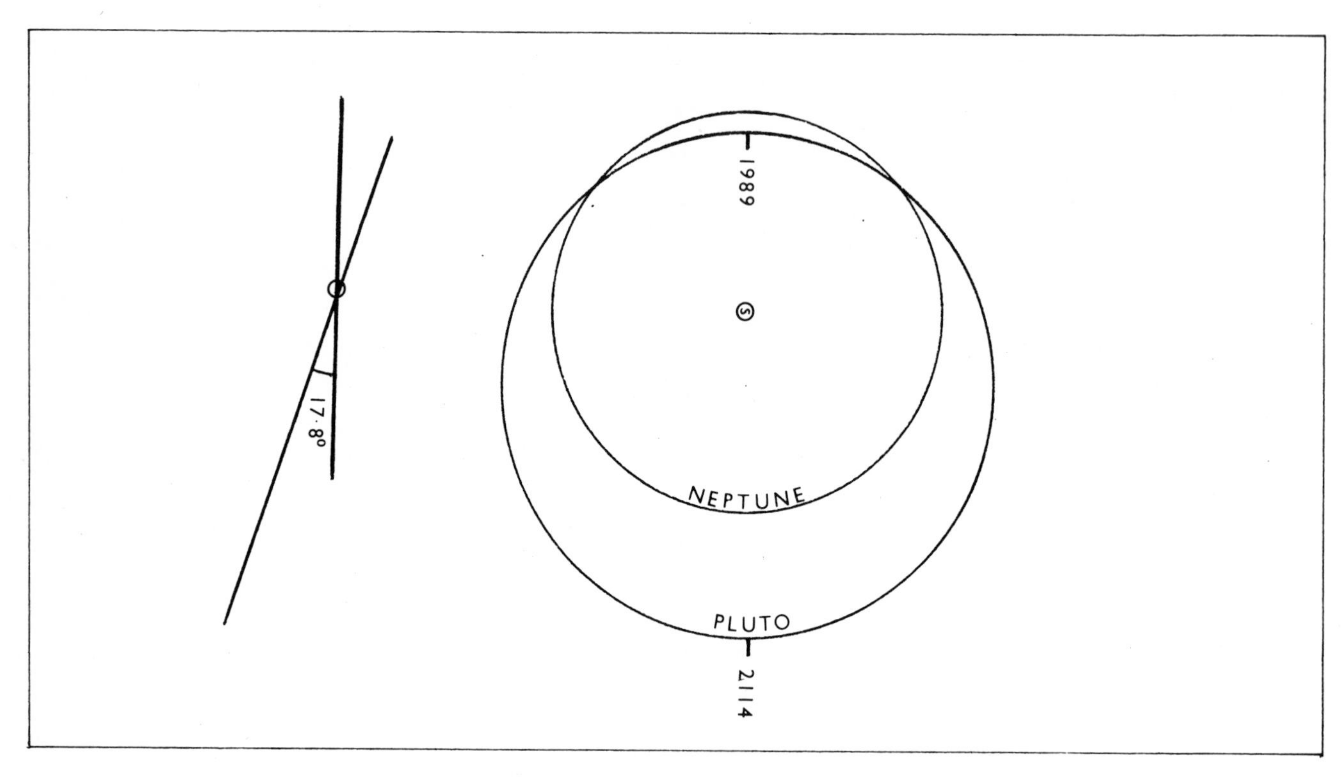

Pluto's orbit crosses that of Neptune so that Pluto is not always the most distant planet from the Sun.

With this discovery the size of the planet has come under review and it seems that the Pluto system comprises a main body, with a diameter of 3,000 kilometres, and a secondary body, with a diameter of 1,200 kilometres and orbiting at a distance of 20,000 kilometres.

The discovery of Pluto has always been something of a mystery. The irregularities in the motion of Uranus were not fully accounted for even with the discovery of Neptune. It was felt at the time that perhaps there was yet another planet beyond the orbit of Neptune, also exerting a pull on Uranus. One astronomer, Percival Lowell, the observer of Martian 'canals', decided to calculate the theoretical position of just such a planet. This he finished in 1914 but it was a while before anything was found. Unfortunately, the eventual discovery of Pluto in 1930 came 15 years too late for Lowell, who died in 1916.

The actual discovery was made photographically by Clyde Tombaugh with the Lowell observatory's 326-mm refractor. Its position agreed well with Lowell's predictions but there was one serious difference. Lowell assumed that a planet exerting the pull necessary to cause the perturbations of the type observed for Uranus must have a mass several times that of the Earth, yet Pluto has a mass less than $\frac{1}{10}$ of the Earth's mass and, therefore, could not account for the perturbations. Was the discovery of Pluto an accident? No one knows. It seems too much of a coincidence and perhaps there is something about the planet that we do not know.

Escaped satellite

As a planet, Pluto is very unlike the gas planets. It has been suggested that it could be an escaped satellite of Neptune. Astronomer-mathematician Fred Hoyle feels that an outer satellite of Neptune approached Triton too closely and the resulting acceleration caused it to break free from its orbit. Possibly as this happened the motion of Triton itself was changed, resulting in its present tetrograde orbit. This would also explain the fact that Pluto, when at perihelion, is closer to the Sun than Neptune. It must be pointed out that when this cross-over actually happens Pluto is not in the same plane as Neptune and the two could never collide.

At the moment Pluto is situated high in the constellation of Virgo and is well placed for northern hemisphere observers. It is, however, moving slowly south and so the conditions for its observation from this point of view are gradually worsening. It must be admitted that there is virtually nothing in the way of useful observation that the amateur can do, but there is an enormous amount of satisfaction to be gained from its location and from watching its slow nightly motion among the stars which, at the time of opposition, amounts to about 1′ each 24 hours. When located it is worthwhile trying to imagine just what the conditions out there must be like.

Whether a spacecraft will ever pay a visit to the planet is anyone's guess. It has been hoped that the Americans could have financed a grand tour of the planets which would have taken advantage of a chance alignment of the planets occurring in the early 1980s but this was not to be. Use is at present being made of this alignment in the form of the Voyager missions and Pluto is not on the itinerary. Obviously someone thinks that it is a long way to go to photograph a ball of ice and rock, so far away from the Sun that the latter looks like nothing more than a very brilliant star.

14.IV.77 21.45 UT and 15.IV.77 22.40 UT 419mm Refl ×248 Only its nightly motion through the stars gives Pluto away.

Occultations and configurations

During the course of their movement through the sky, the planets may, on occasions, either pass in front of a star, pass very close to one another or be occulted themselves by the Moon. Such events are very useful for establishing accurate measurements of the planets and the composition of their atmosphere. Apart from this, such events can be of exceptional beauty.

Perhaps the few illustrations shown here will give some idea of the beauty and fascination constantly found in the planets.

Left *2.III.74 23.45 UT 284mm Refl ×250 Disappearance of Saturn during lunar occultation taking less than 1 minute.* Below left *3.III.74 00.41 UT 254mm Refl ×250 Reappearance of Saturn.* Below *17.VII.74 10h 13m 5s UT 254mm Refl ×280 Lunar occultation of Venus during daylight.* Right *7.IV.76 00.57 UT 254mm Refl ×250 Mars about to occult Epsilon Geminorum.* Far right *4.VI.78 21.20 UT 385mm Refl ×230 Conjunction of Saturn and Mars.*

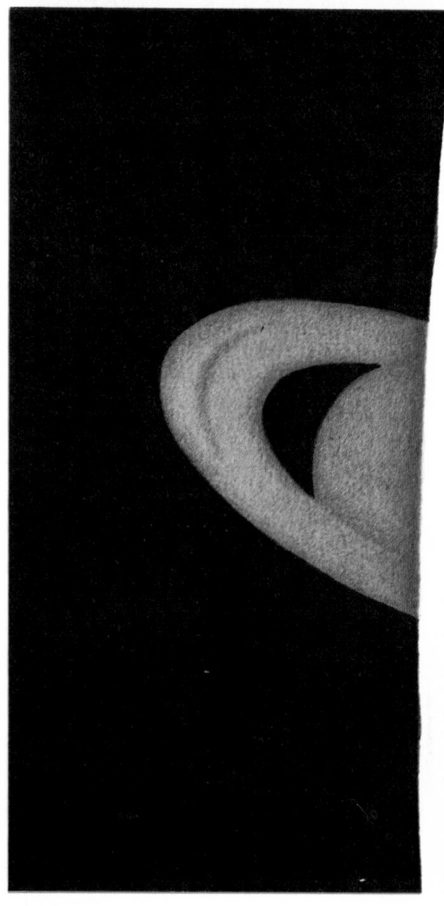

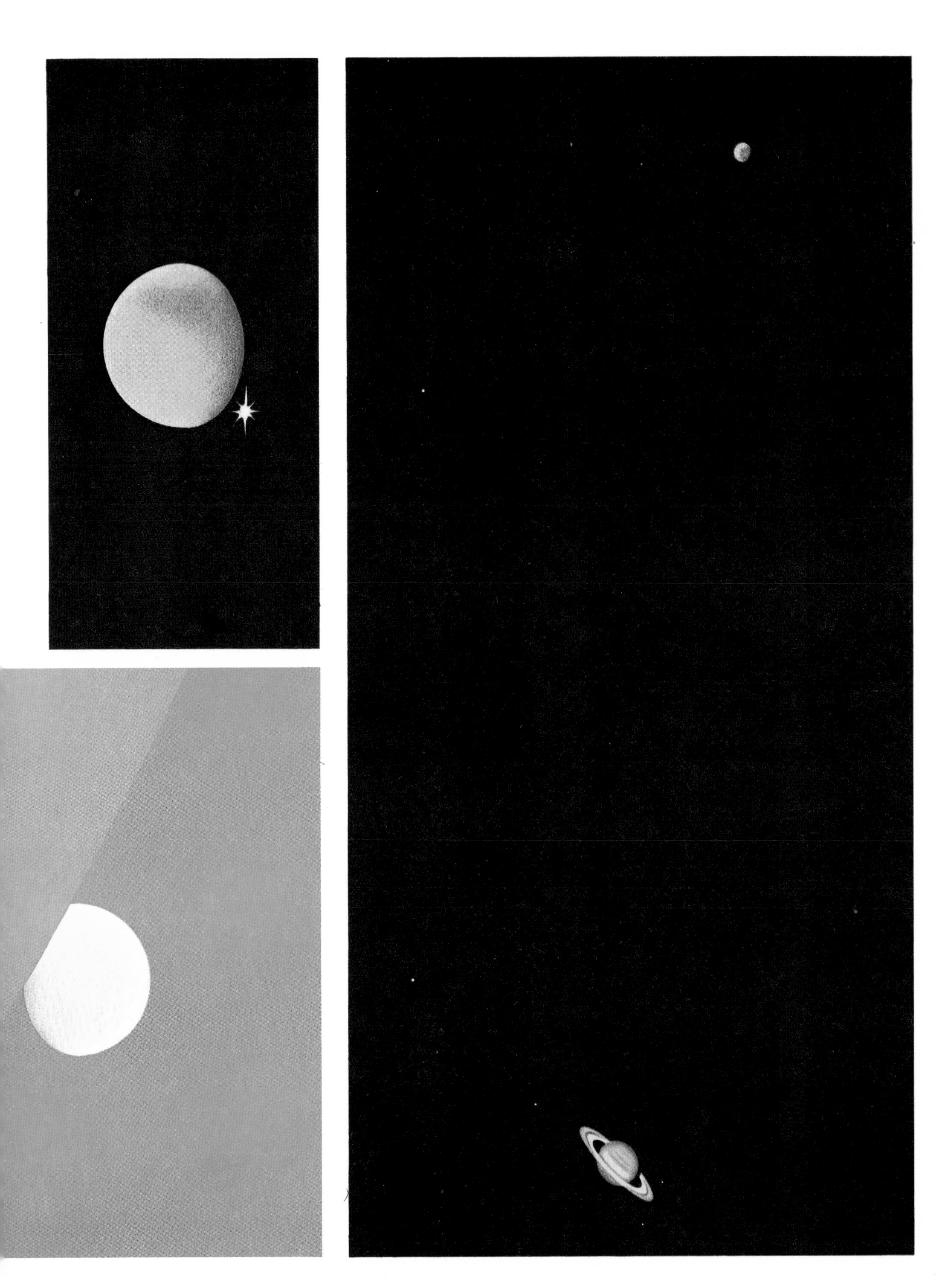

Bibliography

Alexander, A. F. O'D. *The Planet Saturn* Faber & Faber, London 1962.
Alexander, A. F. O'D. *The Planet Uranus* Faber & Faber, London 1965.
Antoniadi, E. M. (Trans. P. Moore) *The Planet Mars* Keith Reid, Devon 1975.
Antoniadi, E. M. (Trans. P. Moore) *The Planet Mercury* Keith Reid, Devon 1974.
Baum, R. M. *The Planets. Some Myths and Realities* David & Charles, Newton Abbot 1973.
Burgess, E. *To the Red Planet* Columbia University Press, New York 1978.
de Callatay, V. & Dollfus, A. *Atlas of the Planets* Heinemann Educational Books, London 1974.
Cross, C. A. & Moore, P. A. *Atlas of Mercury* Mitchell Beazley, London 1977.
Gehrels, T. (Ed.) *Jupiter. Studies of the Interior, Atmosphere, Magnetosphere and Satellites* University of Arizona Press 1976.
Moore, P. A. *Guide to Mars* Lutterworth Press, London 1977.
Moore, P. A. *Guide to the Planets* Lutterworth Press, London 1977.
Moore, P. A. *The Planet Venus* Faber & Faber, London 1960.
Moore, P. A. (Ed.) *Practical Astronomy for Amateurs* Lutterworth Press, London 1963.
Murray, B. *Flight to Mercury* Columbia University Press, New York 1977.
Mutch, T. A. *Geology of Mars* Princeton University Press, New Jersey 1976.
Peek, B. M. *The Planet Jupiter* Faber & Faber, London 1958.
Robinson, J. Hedley *Astronomy Data Book* David & Charles, Newton Abbot 1972.
Robinson, J. Hedley *Using a Telescope* David & Charles, Newton Abbot 1978.
Sandner, Werner *The Planet Mercury* Faber & Faber, London 1963.
Sidgwick, J. B. *Amateur Astronomers' Handbook* Faber & Faber, London 1971.
Sidgwick, J. B. *Observational Astronomy for Amateurs* Faber & Faber, London 1971.

Pages 138–139 *An imaginary view of Saturn—the author's favourite planet—as seen from Titan. The orange smog that hangs in layers in Titan's atmosphere is broken by clear patches, making Saturn plainly visible from its largest moon. This atmosphere is believed to be made up of hydrocarbons, such as methane and acetylene, and the reddish colour can easily be seen by Earth-based observers.*

Although Titan is at a great distance from Saturn, the ringed giant would still appear large in Titan's sky. Also, because Titan's orbit lies in the ring plane of Saturn, the rings would never appear more open than they are here.

This early-morning view from Titan shows Saturn rising ahead of the Sun. Although still below the horizon, the Sun will be shining in the upper layers of Titan's atmosphere, causing the sky brightness shown here. Sometime in the future, a space probe may set down on Titan and be confronted by this impressive spectacle.

Endpapers—back and front *Looking out into the sky from the remote Pluto—an imaginary view. Chiron, the newly discovered satellite of Pluto, would at times be visible and is shown in the foreground. Although much smaller than the planet, Chiron is orbiting so close that it would present an imposing sight.*

In this picture the satellite is seen eclipsing the star-studded Milky Way while the Sun appears as just a very bright, distant star.

Glossary

albedo ratio of the amount of light reflected by a planet in all directions to that which it receives from the Sun.

altitude angular distance of an object above the horizon.

ansae extremities of Saturn's rings as they project from the planet.

aphelion orbital position of a planet when furthest from the Sun.

asteroid minor planet (meaning 'star like').

astronomical unit mean distance of Earth from the Sun; used as a unit of measurement in the Solar System.

atmosphere gaseous envelope surrounding a celestial body.

conjunction apparent close approach of two celestial objects.

constellation star group or star pattern.

convection matter continually agitated by rising hot gas and falling cool gas.

cusps horns or points of a crescent.

declination angular distance of a body north or south of the celestial equator.

density mass per unit volume.

dichotomy exact half-phase of a planet or of the Moon.

eclipse occultation of one celestial object by the shadow of another.

ecliptic apparent path of the Sun through the stars due to Earth's yearly motion around the Sun.

elongation difference between the celestial longitudes of the Sun and a given planet.

first point of Aries Vernal point.

gibbous phase between half and full.

hydrogen lightest and most abundant element.

inclination angle between the plane of a planet's orbit and the plane of the ecliptic.

inferior planet planet whose orbit lies inside the Earth's orbit.

limb actual edge of a planet's disc.

magnitude measurement of the brightness of a star or planet.

nodes points where the orbit of a planet intersects the plane of the ecliptic. The planet is moving from south to north of the plane when in the descending node, north to south in the ascending node.

occultation moon or planet passing between observers and another celestial body. This event causes the disappearance of the second body.

opposition position of a superior planet when in line with the Sun and Earth and diametrically opposite to the Sun.

orbit path followed by a celestial body.

perihelion orbital position of a planet when closest to the Sun.

perturbation small variation in the orbit of a celestial body, arising from the gravitational effect of other bodies.

phase shape of the sunlit portion of a non-luminous celestial body, e.g. crescent, gibbous, full.

reflector telescope that uses a system of mirrors to collect light.

refractor telescope using lenses through which light is refracted.

resolution separating power of a telescope.

revolution orbital motion of a celestial body.

rotation motion of a celestial body as it turns about its axis.

satellite celestial body in orbit around a planet.

scintillation twinkling arising from continual changes in the refractive index of atmospheric layers.

sidereal period time a planet takes to complete one orbit around the Sun.

superior planets planets with orbits outside the Earth's orbit.

synodic period time a planet takes to complete one revolution as seen from the Earth, e.g. time between two successive oppositions.

terminator dividing line between lit and unlit hemispheres of a planet or Moon.

transit inferior planet passing across the disc of the Sun, or satellite passing across the disc of a planet.

occultation Moon or planet passing between observer and another celestial body.

Vernal point point of intersection of the ecliptic with the celestial equator, called the first point of Aries. When the Sun is in this position it is passing from south to north of the equator and thus marks the beginning of Spring.

zenith point on the celestial sphere that is directly over the observer's head.

Zodiac band of sky to either side of the Sun's apparent path around the celestial sphere (Ecliptic).

Index